Full Stack Development with Angular and GraphQL

Learn to build scalable monorepo and a complete
Angular app using Apollo, Lerna, and GraphQL

Ahmed Bouchefra

BIRMINGHAM—MUMBAI

Full Stack Development with Angular and GraphQL

Group Product Manager: Pavan Ramchandani

Publishing Product Manager: Bhavya Rao

Senior Editor: Hayden Edwards

Content Development Editor: Abhishek Jadhav

Technical Editor: Simran Udasi

Copy Editor: Safis Editing

Project Coordinator: Ajesh Devavaram

Proofreader: Safis Editing

Indexer: Sejal Dsilva

Production Designer: Shankar Kalbhor

Marketing Coordinator: Anamika Singh

First published: February 2022

Production reference: 1280222

Published by Packt Publishing Ltd.

Livery Place

35 Livery Street

Birmingham

B3 2PB, UK.

ISBN 978-1-80020-246-7

www.packt.com

To my mother, and to the memory of my father.

– Ahmed Bouchefra

Contributors

About the author

Ahmed Bouchefra is a software developer and technical author with an engineering degree in software development. He is an expert in web development using modern technologies such as Angular and Node.js, along with using traditional technologies such as Python and Django. He's an open sourcer, developer, and maintainer of multiple open source Angular libraries, such as ngx-qrcode2. He also currently writes tutorials about modern web development on techiediaries.com and other industry-leading websites.

About the reviewer

Peter Eijgermans is a long-time software developer and an adventurous and passionate Codesmith frontend developer at Ordina Netherlands. He likes to travel around the world on his bike, always searching for the unexpected and unknown. As part of his job, he tries out the latest techniques and frameworks. He loves to share his experience by speaking at conferences all over the world and writing for the Dutch Java magazine and DZone. He believes that frontend developers are the spiders in the web to bring the user, the project team, and the product together.

I would like to thank the following colleagues for their help in reviewing the book:

Corneel Eijsbouts

Jeroen Verhoeven

Michael Awad

Table of Contents

3

Connecting the Database with TypeORM

4

Implementing Authentication and Image Uploads with Apollo Server

5

Adding Realtime Support with Apollo Server

Part 2: Building the Angular Frontend with Realtime Support

6

Angular Application Architecture and Routing

7

Adding User Search Functionality

8

Guarding Routes and Testing Authentication

9

Uploading Images and Adding Posts

10

Fetching Posts and Adding Comments and Likes

Part 3: Adding Realtime Support

11
Implementing GraphQL Subscriptions

Index

Other Books You May Enjoy

Preface

Angular is one of the most popular JavaScript frameworks in modern web app development, allowing developers to not only build apps and reuse the code, but also develop apps for any deployment target. For web, mobile web, native mobile, and native desktop, GraphQL is the modern and REST alternative for querying web APIs. Using Angular, TypeScript, and GraphQL will give you a future-proof and scalable stack that you can start building apps around.

This book shows you how to build apps using cutting-edge technologies. You'll learn how to solve common web development problems with GraphQL and Apollo, such as database access, authentication, and image uploads.

The book starts by introducing you to building full stack apps with Angular and GraphQL. After that, you'll learn how to create a monorepo project with Lerna and NPM workspaces and configure a Node.js app to use GraphQL with Express and Apollo Server. You'll also understand the basics of Angular architecture and routing. Then, the book demonstrates how to build a professionally looking UI with Angular Material and use Apollo Client to interface with the server to get data from the built-in GraphQL API.

You'll learn about Apollo Client's type and field policies, and various fetching policies. In addition to this, you'll learn about local state management and reactive variables with Apollo Client and how to generate TypeScript types and even Angular Apollo services for your GraphQL schema and queries using GraphQL Code Generator to implement a scalable code base.

By the end of this book, you'll have the skills you need to be able to build your own full stack application.

Who this book is for

This Angular GraphQL book is for Angular developers who want to learn how to use GraphQL and Apollo with Angular to build full stack applications. This book does not assume prior knowledge of full stack development with Angular and GraphQL.

What this book covers

Chapter 1, App Architecture and Development Environment, teaches you about the project's structure and the tools required to develop the application. After that, you'll have your machine ready for development. Specifically, you'll install Node.js alongside npm and MySQL.

Chapter 2, Setting Up GraphQL with Node.js, Express.js, and Apollo, assists you with building your backend application to provide the API that will be consumed by your Angular application.

Chapter 3, Connecting the Database with TypeORM, teaches you how to connect a MySQL database to your application using TypeORM, as well as how to create resolvers to get and save data from the database.

Chapter 4, Implementing Authentication and Image Uploads with Apollo Server, helps you understand how to add authentication and image uploads with Apollo Server to your GraphQL API, implementing more resolvers. You'll learn about the necessary concepts for adding authentication with Node.js, Express, and Apollo Server and then how to handle image uploads.

Chapter 5, Adding Realtime Support with Apollo Server, helps you to add realtime support to your server application, which will allow you to communicate fresh data from the server to the client as soon as it becomes available. To do this, you'll use Apollo Server's GraphQL subscriptions.

Chapter 6, Angular Application Architecture and Routing, gets you started by installing the Angular CLI and creating a new project using a recent version of Angular. Following that, you'll utilize the Angular CLI to create the modules, services, and components that make up your application's UI, as well as being introduced to dependency injection.

Chapter 7, Adding User Search Functionality, specifically looks at how to integrate the frontend with the backend using Apollo Client, which is designed for sending GraphQL queries and mutations to the server to fetch and write data. Then, you'll begin implementing authentication.

Chapter 8, Guarding Routes and Testing Authentication, continues with implementing our authentication system by guarding the necessary route(s) from unauthorized access, sending the JWT with API requests, and unit testing our code.

Chapter 9, Uploading Images and Adding Posts, implements the profile component's functionality. You'll add the necessary code to fetch the user that corresponds to a profile URL and render their information on the page, including the ability to upload the user's photo and cover image as well as add a biography.

Chapter 10, Fetching Posts and Adding Comments and Likes, begins by dealing with an authentication token expiration issue, and then you'll continue working on your profile component by sending queries to receive paginated posts and comments data and mutations to add comments and likes to posts.

Chapter 11, Implementing GraphQL Subscriptions, continues with building your users' profile component before learning how to add realtime support to your application so that you can retrieve and display new data from the server without having to constantly refresh the app. You'll utilize GraphQL subscriptions with Apollo Client and Angular to do this.

To get the most out of this book

Software/hardware covered in the book	Operating system requirements
Angular 12.2.8	Ubuntu
Apollo-Server-Express: 3.3.0	
Apollo Client 3.4.16	
TypeScript 4.3.5	
Node 12.22.1	

Installing Node.js and MySQL is covered in the book.

Further prerequisites are being familiar with JavaScript/TypeScript, HTML, and CSS.

If you are using the digital version of this book, we advise you to type the code yourself or access the code from the book's GitHub repository (a link is available in the next section). Doing so will help you avoid any potential errors related to the copying and pasting of code.

Download the example code files

You can download the example code files for this book from GitHub at `https://github.com/PacktPublishing/Full-Stack-App-Development-with-Angular-and-GraphQL`. If there's an update to the code, it will be updated in the GitHub repository.

We also have other code bundles from our rich catalog of books and videos available at `https://github.com/PacktPublishing/`. Check them out!

Download the color images

We also provide a PDF file that has color images of the screenshots and diagrams used in this book. You can download it here: `https://static.packt-cdn.com/downloads/9781800202467_ColorImages.pdf`.

Conventions used

There are a number of text conventions used throughout this book.

`Code in text`: Indicates code words in text, database table names, folder names, filenames, file extensions, pathnames, dummy URLs, user input, and Twitter handles. Here is an example: "Open the `lerna.json` file and set `version` to `"0.0.1"`:"

A block of code is set as follows:

```
{
  "packages": [
    "packages/*"
  ],
  "version": "0.0.1"
}
```

When we wish to draw your attention to a particular part of a code block, the relevant lines or items are set in bold:

```
import express, { Application } from 'express';
import { ApolloServer } from 'apollo-server-express';
import schema from './graphql/schema';
```

Any command-line input or output is written as follows:

```
curl -sL https://raw.githubusercontent.com/nvm-sh/nvm/v0.37.2/install.sh
```

Bold: Indicates a new term, an important word, or words that you see onscreen. For instance, words in menus or dialog boxes appear in **bold**. Here is an example: "You can click on the **Query your server** button to access Apollo Studio, where you can send requests to your API endpoint."

> Tips or Important Notes
> Appear like this.

Get in touch

Feedback from our readers is always welcome.

General feedback: If you have questions about any aspect of this book, email us at customercare@packtpub.com and mention the book title in the subject of your message.

Errata: Although we have taken every care to ensure the accuracy of our content, mistakes do happen. If you have found a mistake in this book, we would be grateful if you would report this to us. Please visit www.packtpub.com/support/errata and fill in the form.

Piracy: If you come across any illegal copies of our works in any form on the internet, we would be grateful if you would provide us with the location address or website name. Please contact us at copyright@packt.com with a link to the material.

If you are interested in becoming an author: If there is a topic that you have expertise in and you are interested in either writing or contributing to a book, please visit authors.packtpub.com.

Share Your Thoughts

Once you've read *Full-Stack Development with Angular and GraphQL*, we'd love to hear your thoughts! Scan the QR code below to go straight to the Amazon review page for this book and share your feedback.

https://www.amazon.in/review/create-review/error?asin=1800202466

Your review is important to us and the tech community and will help us make sure we're delivering excellent quality content.

Part 1: Setting Up the Development Environment, GraphQL Server, and Database

In this part, we'll learn about our app's architecture and requirements, and we'll set up our development environment for full stack development. We'll also set up a GraphQL server with Express.js and Apollo and connect a SQL database for storing the application data using TypeORM.

This section comprises the following chapters:

- *Chapter 1, App Architecture and Development Environment*
- *Chapter 2, Setting Up GraphQL with Node.js, Express.js, and Apollo*
- *Chapter 3, Connecting the Database with TypeORM*
- *Chapter 4, Implementing Authentication and Image Uploads with Apollo Server*
- *Chapter 5, Adding Realtime Support with Apollo Server*

1
App Architecture and Development Environment

Throughout this book, we'll develop a full stack web application with an **Angular** frontend, a **Node.js** backend, and a **MySQL** database.

As a consequence, in this first chapter, we'll learn about the project's structure and the tools required to develop the application. After that, we'll have our machine ready for development. Specifically, we'll install Node.js alongside npm and MySQL.

Node.js is required for the backend application, which runs a server that exposes a GraphQL API using Express.js, one of the most popular Node frameworks. Node.js is also necessary for development of the frontend application. This is because Angular uses an official command-line interface to initialize the project and scaffold any required artifacts during the application's development.

MySQL will be utilized to store our data but it will not be directly accessible in our code. Rather than that, we will be using an ORM called **TypeORM**.

In this chapter, we will cover the following topics:

- The project's architecture and development technologies
- Installing MySQL
- Installing and configuring Node.js

Technical requirements

This chapter will require access to a computer equipped with an operating system and an internet connection. Because I'll be working with an Ubuntu system, the instructions in this book will be specific to that system. They should also function on any other system that is based on Debian.

The architecture and technologies

In this section, we'll learn about the architecture of our application and the technologies that we'll use to develop both the frontend and backend of our application throughout this book.

Let's begin by familiarizing ourselves with the notions of **full stack** architecture and **monorepo** repositories (also known as **mono-repositories**).

Full-stack architecture

We'll be developing a web application that includes both a frontend and a backend, also known as a **full stack application**. This implies that we will take the job of a full stack developer.

A full stack developer is a web developer who can handle the development of both the frontend (client-side) and backend (server-side) parts of a web development project.

They must be proficient in fundamental technologies such as HTML, JavaScript, and CSS and have some familiarity with others such as TypeScript, Node.js, and database management systems such as MySQL. These technologies may be broadly classified into two types:

- Client-side languages, frameworks, and tools for developing browser-based applications, including HTML, CSS, JavaScript, TypeScript, and Angular.
- Languages and tools for the server side, such as Node.js, Python, or PHP, as well as database languages, such as SQL.

To build our full stack application, we'll need two parts:

- **The frontend**: This is the application that runs in the client's web browser on the client's machine. Historically, this was accomplished through the use of HTML pages rendered on the server and then returned to the client. However, browsers may now run fully fledged JavaScript applications (also known as **client-side apps**), which do the majority of the processing on the client's browser and rely on servers just to supply the application's initial files and data. Simply put, the frontend is what presents the user interface with which users interact.

- **The backend**: This is the server-side application that will handle HTTP requests, perform some processing logic, and return responses to the browser (HTML and/or JSON data).

We'll use tools such as Lerna to organize our application's code utilizing a monorepo approach. In the context of software development, monorepo simply refers to using a single source code repository (typically version controlled using Git) for all of our applications (the server, client, and any shared libraries).

Let's take a high-level look at how our application will be delivered to the client's machine:

1. First, the client will make a request by putting your application domain name into the address bar of the browser.

2. The server will intercept the requests and process the HTML document containing the Angular app.

3. The client's browser will begin downloading all of the JavaScript and CSS files required to run the Angular application.

4. Any initial requests to the server for data, such as posts, will be sent by the Angular application and rendered in the user interface.

5. The other requests will be made to the server after the user begins engaging with the application.

This process provides a high-level overview of how our application operates. We'll go through these stages in greater depth in the next chapters.

After quickly discussing the full stack and monorepo concepts, let's have a look at the technologies and tools we'll be using to build our application.

The development technologies

Every project necessitates the use of a set of technologies and tools. This includes programming languages, command-line interfaces, libraries, and frameworks. We'll be using diverse technologies for both frontend and backend development in our full stack project.

Nowadays, modern development involves using the same tools for both the frontend and the backend. Let's look at these technologies and tools in more detail.

We'll use a recent version of Angular (version 12 at the time of writing) for the frontend, Node.js for the backend, and a set of supporting libraries including Express.js for launching a web server and TypeORM for abstracting database operations.

One of the primary goals of this book is to build a **GraphQL** API that will be provided by our Express.js server and consumed by our Angular frontend. To abstract all of the low-level APIs necessary to run a web server, we'll utilize Express.js on top of Node.js.

The web server will listen for incoming HTTP requests from the client's browser, which are mostly GraphQL queries for getting data and mutations for creating and updating data.

Following that, we'll look at Node.js and what it is used for.

Running JavaScript on the server with Node.js

What exactly is Node.js? Node.js is a free and open source platform and runtime environment that allows you to run JavaScript on your server and employs an event-driven, non-blocking input/output (I/O) architecture. Ryan Dahl created Node.js in 2009 on top of Google Chrome's JavaScript Engine (V8 Engine).

If you're a seasoned JavaScript developer, you're probably familiar with JavaScript as a programming language used to create dynamic web pages that can only be executed in the client's browser. However, thanks to Node.js, we can now utilize JavaScript to build web apps on both the client (through the browser) and the server.

As you might expect, developers may now utilize a single programming language to create their complete full stack web application rather than utilizing a distinct language for the server, such as PHP or Python, which are just two of the many available alternatives. Node.js, like these languages, may be used to develop the backend of your web applications.

This enables frontend JavaScript developers to begin developing backend apps without learning a new language.

Node.js is used for more than just server-side applications; it's a general-purpose tool for building all kinds of network apps, as well as for building and running frontend desktop tools on the developer's machine. The Angular CLI, for example, is a Node.js-based tool that we'll use to develop and work with our frontend Angular project.

> **Important Note**
> Because Node.js is a JavaScript runtime, it may also be used with TypeScript, which is a JavaScript superset that includes object-oriented programming principles and static typing. In this book, we'll develop our backend utilizing Node.js and TypeScript.

Installing packages with npm

Node.js has a large ecosystem as well as a package management tool known as Node Package Manager (npm). It may be used to quickly install packages from a central registry containing thousands of packages built and published by other organizations and developers to tackle common development problems.

You don't have to reinvent the wheel while trying to address the same development difficulties that other developers have previously faced, thanks to the large ecosystem and the npm registry. The npm registry also makes it straightforward to install any package or library with a single command, including well-known libraries such as Angular and Express.js.

On both our frontend and backend projects, we'll use npm to install the necessary tools and libraries, such as Angular, Apollo, and Express.js, among others.

Running a web server with Express.js

As previously stated, Node.js is a platform and runtime environment that exposes a set of low-level APIs that we rarely use directly while building web apps. Instead of developing a lot of sophisticated code or reinventing the wheel, we'll leverage certain libraries created by other developers. Express.js, as we previously stated, is one of these libraries. So, what exactly is it?

Express.js is a popular Node.js web application framework for developing server applications. It's unopinionated, lightweight, and includes all of the fundamental capabilities needed to develop web applications and web APIs.

Express.js saves you from having to deal with low-level Node.js APIs to create servers that receive HTTP requests and respond with HTTP responses. It also makes it simple to implement routing, manage static files, and serve assets.

> **Tip**
>
> Unlike popular frameworks such as Django in Python, which is considered opinionated, Express.js is a flexible framework that does not enforce how you should organize your project.

GraphQL

Now that we know what Node is and that we'll use Express.js to run our backend server, what is GraphQL and where does it fit in this stack?

GraphQL is similar to SQL; however, it is used to query web APIs rather than databases. It is an API specification and query language. It also serves as a runtime for responding to queries in order to get, create, and update data.

> **Important Note**
>
> GraphQL is a newer alternative to REST, which is widely used by developers to create APIs that can be consumed by both desktop and mobile clients.

Facebook introduced GraphQL in 2015 to address some of the shortcomings of REST and other API development methodologies.

For example, with GraphQL, the frontend application may request only the data it requires from the server. This is due to the fact that GraphQL allows you to describe the forms of your data using user-defined types that are formed from fundamental built-in types such as strings and integers.

GraphQL may be used with a variety of network protocols, the most prevalent of which are WebSocket and HTTP.

If you're acquainted with SQL, you'll recognize that it's comparable to how you define the form of your data using tables. If you're acquainted with object-oriented languages, it's also comparable to how you build interfaces and classes to describe real-world objects.

You can send queries that are JSON-like objects that specify the fields you want the server to return with an HTTP response. The following is an example of a query:

```
post {
    id
    content
    date
}
```

If this query is submitted to a GraphQL server with a post type defined with the ID, content, and date attributes, the associated post data will be returned using **resolver functions**.

A resolver function handles the resolution for data, executes the logic necessary to retrieve data from the database, and returns it to the requesting client.

As previously stated, GraphQL is a standard that is not bound to any programming language or framework. Many popular programming languages, such as Python and JavaScript, have implementations.

It's also not bound to any database system and may be used with any technology stack that includes MySQL, or any database management system, as a database. In our situation, we'll be utilizing it in conjunction with MySQL.

Apollo is one of the GraphQL implementations. So, where does it fit in our technology stack?

Integrating the frontend and backend with Apollo

Apollo, an industry-standard GraphQL implementation for JavaScript, will be used. It consists of a client and a server part, referred to as Apollo Client and Apollo Server, respectively.

Apollo Client is compatible with plain JavaScript as well as recent UI libraries and frameworks (such as React, Angular, and Vue.js), while Apollo Server is compatible with common Node.js frameworks (such as Express.js and Hapi).

We can simply and effortlessly interact between the frontend and backend of our application thanks to Apollo Client and Apollo Server, which eliminates the need for complex data fetching logic.

So, in our Angular frontend, the Apollo Client will provide a layer for sending queries to obtain data as well as mutations for adding, modifying, and removing data from the database. The Apollo Client does not interface with the database directly, but rather with the Apollo Server and Express.js, which operate on top of Node.js.

Saving data with TypeORM and MySQL

The data will be saved in a MySQL database. In development, we'll use a locally installed MySQL server, but in production, you may use a cloud-hosted relational database such as **Amazon Relational Database Service** (**Amazon RDS**) or any other service of your choosing.

We picked MySQL as our database management system since it is the most widely used open source relational database in the world, meaning most developers are acquainted with it and may have used it previously in one of their web projects.

It is very easy to install locally on all supported operating systems. It's also simple to set up and scale in production thanks to cloud services such as Amazon RDS and DigitalOcean.

Amazon RDS offers a free tier, and after that is reached, you will only be charged following a pay-as-you-go basis. It will help you to focus on application development rather than database management operations such as backups, monitoring, scalability, and replication.

We will not be using SQL to build tables or query data directly. Instead, we'll use an Object Relational Mapper (ORM) to create database tables and query, insert, and remove data using a high-level programming language rather than SQL. TypeORM, a TypeScript-based ORM, will be used in our case.

Now that we've covered the final major component of our backend, the database, let's have a look at another component of our technological stack: Angular.

Building the frontend with TypeScript and Angular

Google's Angular is an open source client-side framework. It was created from the ground up in TypeScript as a replacement for Angular.js, which was based on plain JavaScript. Angular, along with React and Vue, is one of the three most popular frontend frameworks. It includes the libraries required to build modern frontend web apps for mobile and desktop devices.

We'll create our project with a basic file structure using the official Angular CLI, and we'll organize our client-side TypeScript code using abstractions such as modules, components, and services.

Angular has a client-side router out of the box, allowing us to add routing and navigation to our application. When we start employing these techniques, we'll go over them in greater depth in the coming chapters.

We'll also be integrating our frontend with the GraphQL server, which is built on Apollo Server, by leveraging Apollo Client with Angular.

We will have a technology stack that includes Node.js, MySQL, Express.js, TypeORM, Apollo, and Angular by joining all of these tools.

Now that we've covered the application architecture and technologies, let's get started by installing MySQL and Node.js in our development environment.

Installing MySQL

In this section, we'll learn how to install MySQL on our development machine. The instructions for installing MySQL on your computer depend on your operating system, but here, we'll focus on the instructions for Ubuntu.

MySQL is a popular database management system that is also useful for local development since it is simple to install and configure.

The installation process is straightforward; simply update your system's package index, install the `mysql-server` package, and then execute the accompanying security script.

MySQL is probably already installed on your development machine. If that's the case, you may skip this step.

> **Important Note**
>
> The steps below are exclusively for setting up MySQL on your local machine for development purposes. In production, you must follow the appropriate guides, especially when it comes to securing your database from attacks. You can easily achieve this with cloud services, which provide a managed database that you don't have to manage or secure yourself.

Now, let's get started by running the instructions to install MySQL Server. Open a new command-line interface and run the following commands:

```
sudo apt-get update
sudo apt-get upgrade -y
```

These instructions will update your system's package index to the most recent version.

Then, to install MySQL Server, use the following command:

```
sudo apt-get install mysql-server
```

Now that we've installed MySQL Server, let's learn how to configure it.

Configuring MySQL Server

After installing MySQL Server, you must execute a security script. Return to the command-line interface and execute the following command:

```
sudo mysql_secure_installation
```

You'll be prompted for your root user password; enter it and press *Enter*.

Next, you'll be presented with a bunch of questions. The first question will be `Would you like to set up VALIDATE PASSWORD plugin?` This is used to validate passwords and increase security. It evaluates the strength of the password and helps users to create passwords that are sufficiently safe. Because we're on a development machine, this isn't critical, therefore answer with *N* for no. The following question will ask you to create a password for the MySQL root user. Enter a password of your choice and confirm it.

To select the default answers for the following questions, just type *Y* and then press *Enter*:

- `Remove anonymous users?`
- `Disallow root login remotely?`
- `Remove test database and access to it?`
- `Reload privilege tables now?`

This will remove some anonymous users as well as a test database and access to it, disable remote root login, and load the privilege tables to guarantee that all previous modifications take effect instantly.

That's all there is to it — you're done! You have successfully installed MySQL Server on your local development machine, which is running Ubuntu. Following that, you'll learn how to verify whether MySQL is running and how to start it if it isn't.

Testing MySQL Server

MySQL should have started automatically after installation. Return to the command-line interface and execute the following command:

```
systemctl status mysql.service
```

If it's up and running, you'll see something like this:

```
  mysql.service - MySQL Community Server
     Loaded: loaded (/lib/systemd/system/mysql.service; enabled;
vendor preset: en
     Active: active (running) since Tue 2020-12-08 17:15:40 +01;
48min ago
Main PID: 20416 (mysqld)
      Tasks: 29
     CGroup: /system.slice/mysql.service
             └─20416 /usr/sbin/mysqld --daemonize --pid-file=/
run/mysqld/mysqld.pi
```

If MySQL Server is not running, use the following command to start it:

```
sudo systemctl start mysql
```

After we've installed MySQL and verified that it's up and running, we'll learn how to install Node.js, which will be required by both our frontend and backend apps.

Installing and configuring Node.js

We'll need to install Node.js after installing the MySQL database management system because it's necessary for executing our server code.

We have several options for installing Node.js on our operating system:

- Node Version Manager (NVM), which you can use to run various versions of Node. js on your development system, acquire information about the available versions, and install any version with a single command

- The operating system's official package manager, such as APT for Ubuntu, Homebrew for macOS, or Chocolatey for Windows

- The binaries from the official website at `https://nodejs.org/en/ download/`, which not only provides Windows, macOS, and Linux binaries, but also source code that can be downloaded and compiled

As previously said, we will presume you are running a Debian-based system such as Ubuntu. In this chapter, we'll teach you how to use the first method on an Ubuntu system.

> **Important Note**
> If you are not using an Ubuntu or Debian-based system, go to the official website at `https://nodejs.org/en/download/package-manager/` and get the necessary instructions to install Node.js on your operating system.

Installing Node.js with nvm

You can install Node.js and npm using nvm instead of your system's native package manager. This utility does not operate at the system level. Instead, it makes use of a distinct folder in your home directory.

This allows you to install several versions of Node.js at the same time and quickly switch between them as needed. In addition, after you've installed nvm, you can quickly install any version of Node.js, old or new, with a single command.

Let's go over the steps:

1. Return to your command-line interface and run the following command to download the nvm installation script:

    ```
    curl -sL https://raw.githubusercontent.com/nvm-sh/nvm/
    v0.37.2/install.sh
    ```

2. Then, use the following command to execute the script:

    ```
    bash install.sh
    ```

    This command will clone the nvm repository under the `~/.nvm` folder and update the `~/.profile` file as necessary.

3. To begin utilizing nvm, just source the following file or log out and then log back in:

    ```
    source ~/.profile
    ```

4. Using this command, you can quickly retrieve a list of available Node.js versions that you can install:

    ```
    nvm ls-remote
    ```

5. We are using `v12.22.1` in this book, which you can install using the following command:

    ```
    nvm install v12.22.1
    ```

    This will download the installation binary that is compatible with your operating system and install Node.js together with npm. It will also make this version the default version.

6. To verify the installed version, use the following command:

    ```
    node -v
    ```

    In our case, we get `v12.22.1` printed on the Terminal.

More information on the available commands and how to use them may be found in the official repository at `https://github.com/nvm-sh/nvm`.

The nvm utility is compatible with Linux, macOS, and **Windows Subsystem for Linux (WSL)**. You may use two unofficial alternatives for Windows:

- nvm-windows, which may be found at `https://github.com/coreybutler/nvm-windows`

- nodist, which may be found at `https://github.com/marcelklehr/nodist`

That's everything – you've configured your development machine to run Node.js, allowing you to install and use Node.js packages such as Express.js to build your backend application, as well as frontend libraries and tools such as Angular CLI.

We'll use npm to install the dependencies for our backend and frontend apps throughout this book.

Summary

In this chapter, we learned about the full stack and monorepo architectures of the application we'll be building throughout this book, as well as the technologies we'll be utilizing to build it.

After covering the architecture and technologies, we looked at how to install MySQL on our development machine. Finally, we learned how to install and set up Node.js, which is required by our full stack application's frontend and backend.

This chapter is now complete! We need to build a Node.js server with GraphQL support, to implement the backend, now that we've set up the development environment and installed Node.js. In the following chapter, we'll begin by installing and configuring Express.js, before adding Apollo Server and looking at how to test and debug our GraphQL server.

2
Setting Up GraphQL with Node.js, Express.js, and Apollo

In the previous chapter, we prepared our development machine by installing **Node.js** and **MySQL**. We can now start building our backend application to provide the API that will be consumed by our **Angular** application.

For the sake of time, we'll try to build a simple social network application with a minimal set of features (a **Minimum Viable Product** or **MVP**) that will be implemented throughout the book.

First, users will be presented with a login and signup interface. In this interface, users are required to provide their email and password to log in or create an account if they are not already registered. In this case, they need to enter their full name, username, email, password, and password confirmation.

After they log in, they will be taken to an interface where they can create posts and see their feed, which will display the posts that have been added by the users of the app. They can also search for the other users by name and visit their profiles from there.

Users can see their profiles, which contain their photos and posts, and the profiles of other users.

Finally, users can add comments and likes on posts and receive notifications when someone comments or likes their posts.

After setting up the development environment and installing Node.js, we'll need to set up a Node.js server with **GraphQL** support to implement the backend. In this chapter, we'll explain how to install **Express.js** and configure it with **TypeScript** and GraphQL.

Next, we'll learn how to use **mocking** to provide a working GraphQL server with **Apollo Server** for testing before implementing the **resolvers**, which are responsible for fetching and mutating data.

We'll also learn how to configure **cross-origin resource sharing** (**CORS**) and learn how to test our GraphQL server with Apollo Studio.

In this chapter, we will cover the following topics:

- Initializing our project
- Installing Express.js and adding TypeScript support
- Setting up Apollo Server with Node.js
- Mocking our GraphQL API with Apollo Server
- Testing the GraphQL server using Apollo Studio
- Configuring CORS in Express.js

Technical requirements

To complete this chapter, you are required to have Node.js and npm installed on your local development machine. Please refer to *Chapter 1, App Architecture and Development Environment*, for instructions on how to install them if you haven't done so yet.

You also need to be familiar with the following technologies:

- JavaScript/TypeScript
- Git, Node.js, and Express
- GraphQL concepts such as schemas and types

If you need a refresher of GraphQL concepts, check out `https://graphql.org/learn/`. This is the official documentation for GraphQL schemas and it explains various features of the schema and how to use them with the schema language.

You can find the complete source code for this chapter at `https://github.com/PacktPublishing/Full-Stack-App-Development-with-Angular-and-GraphQL/tree/main/Chapter02`. Alternatively, you can go to `https://git.io/JKZpT`. Make sure that you consult the history, which contains the commits for the major steps in this chapter.

Setting up a monorepo project

We'll be taking a monorepo approach to managing our project, which simply revolves around the idea of using a single repository to manage our client and server apps (and any shared libraries) instead of working with multiple repositories. We'll be using a tool called **Lerna** to do this. Let's get started:

1. Head over to your command-line interface and run the following command to install Lerna:

   ```
   npm install --global lerna
   ```

2. Next, create a folder for your project and initialize Lerna using the following commands:

   ```
   mkdir ngsocial
   cd ngsocial
   lerna init
   ```

   You should see an output similar to the following:

   ```
   lerna notice cli v4.0.0
   lerna info Initializing Git repository
   lerna info Creating package.json
   lerna info Creating lerna.json
   lerna info Creating packages directory
   lerna success Initialized Lerna files
   ```

   The `lerna.json` file holds the configuration for Lerna, while the `package.json` file contains the configuration for the whole (client and server) project. The `packages/` folder will contain the client and server apps, plus any libraries you would like to share between the two apps. The command will also run the `git init` command to initialize a Git repository. You can find the available options for configuring Lerna at `https://github.com/lerna/lerna#lernajson`.

3. Next, create a `.gitignore` file and add the following contents:

```
node_modules
.DS_Store
```

4. Next, open the `lerna.json` file and set `version` to `"0.0.1"`:

```
{
  "packages": [
    "packages/*"
  ],
  "version": "0.0.1"
}
```

Then, open the `package.json` file and set `version` to `"0.0.1"`:

```
{
  "name": "root",
  "private": true,
  "devDependencies": {
    "lerna": "^4.0.0"
  },
  "version": "0.0.1"
}
```

For more information on Lerna, check out `https://lerna.js.org/`.

5. Next, head over to your GitHub account and create a new repository that you can push your project's code to. This process is simple, but if you encounter any issues, make sure that you consult the docs at `https://help.github.com/articles/creating-a-new-repository/`.

6. Then, you need to point your local repository to your remote repository using the following command:

```
git remote add origin <PUT_YOUR_REPO_URL_HERE>
```

7. Before progressing, let's create a branch so that we can differentiate between the code for each chapter. In your Terminal, run the following command:

```
git checkout -b chapter2
```

You should get the following output:

```
Switched to a new branch 'chapter2'
```

8. Finally, stage and commit the changes using the following command:

```
git add -A
git commit -m  "Init lerna"
```

9. Then, if you want to push the code, run the following command:

```
git push -u origin chapter2
```

Now that we have created a monorepo project using Lerna, let's initialize our server-side project.

Initializing our server project

In this section, we'll learn how to initialize our server project and install Express.js, which will help us quickly spin up a server without dealing with the low-level APIs provided by the Node.js platform.

npm provides a command that allows developers to create a project (an empty folder with a package.json file). Let's learn how to use it to generate a package.json file.

Generating a package.json file

We'll get started by creating a folder for our server files inside the ngsocial/ packages/ folder:

1. Open a new command-line interface and run the following commands:

```
cd packages
mkdir server && cd server
```

2. Next, run the following command:

```
npm init --yes
```

The --yes flag tells npm to generate a package.json file with the default settings.

3. You can run the following command to update your npm version:

```
npm install -g npm
```

We are executing this command because we need to update to npm 7 (if you are not already using this version, which includes support for **Workspaces**). See https:// docs.npmjs.com/cli/v7/using-npm/workspaces.

The following is the content of the package.json file. Here, in my case, I have tweaked the name, author, version, and license information:

```
{
  "name": "ngsocial",
  "version": "0.0.1",
  "description": "",
  "main": "index.js",
  "scripts": {
    "test": "echo \"Error: no test specified\" && exit 1"
  },
  "keywords": [],
  "author":
    "Ahmed Bouchefra (https://www.ahmedbouchefra.com)",
  "license": "MIT"
}
```

Now that we have a project folder with a package.json file, next, we'll install Express.js and some development dependencies such as typescript and ts-node.

Installing Express.js and development dependencies

After initializing our server project, we need to install the express package from the npm registry using the following command:

```
npm install express
```

At the time of writing, express v4.17.1 is installed. In the package.json file, a new object named dependencies must be added, as follows:

```
"dependencies": {
  "express": "^4.17.1"
}
```

Next, we need to install the following development dependencies:

- typescript, a tool that enables you to compile the TypeScript code to plain JavaScript
- ts-node, a tool that enables you to start a development server based on TypeScript directly from the Terminal without compiling it to plain JavaScript

Head back to your Terminal and run the following command:

```
npm install --save-dev typescript ts-node
```

After installing these libraries, open the package.json file. You'll see a new devDependencies object, as follows:

```
"devDependencies": {
   "ts-node": "^10.2.1",
    "typescript": "^4.4.3"
}
```

We'll also have a new executable called tsc that refers to the TypeScript compiler. Since we have installed the package locally, we need to provide the full path to invoke it. For example, we can get the installed version of typescript by providing the full path of the executable with the –v switch, as follows:

```
./node_modules/typescript/bin/tsc -v
Version 4.4.3
```

Now, let's use this tool to generate a tsconfig.json file in the root of our server folder:

```
./node_modules/typescript/bin/tsc --init
message TS6071: Successfully created a tsconfig.json file.
```

> **Important Note**
> We can also use the npx tsc --init --rootDir src
> --outDir dist --lib dom,es6 --module commonjs –
> removeComments command to generate a tsconfig.json file with
> only the properties that we need. Please refer to https://aka.ms/
> tsconfig.json to read more about the tsconfig.json file.

The tsconfig.json file contains all the possible options for configuring TypeScript. Let's remove most of them and only keep the following minimal configuration:

```
{
   "compilerOptions": {
     "target": "es6",
     "module": "commonjs",
     "rootDir": "./",
     "outDir": "./dist",
```

```
    "esModuleInterop": true,
    "strict": true
  }
}
```

These are the options that we have left in the file:

- `target`: This is used to specify the JavaScript version of the compiler's output.
- `module`: This is used to specify the module system of the compiled JavaScript code.
- `rootDir`: This is used to specify where the TypeScript code files are located in your project.
- `outDir`: This is used to specify where the output folder of the compiled JavaScript code is located.
- `esModuleInterop`: This tells the compiler to compile ES6 modules to CommonJS modules.
- `strict`: This is used to enable strict type-checking.

Next, we need to install the `type` definitions for Node and Express.js using the following command:

```
npm install –save-dev @types/node @types/express
```

They will be added under the `devDependencies` object in the `package.json` file:

```
"devDependencies": {
  "@types/express": "^4.17.13",
  "@types/node": "^16.9.6",
  "ts-node": "^10.2.1",
  "typescript": "^4.4.3"
}
```

Optionally, you can stage and commit the changes, as follows:

```
git add -A
git commit -m "Installing Express and development dependencies"
```

After installing Express.js and the other development dependencies and configuring TypeScript in our project, let's learn how to create the server.

Creating the server

Now, we're ready to create our Express server:

1. Create a `src/index.ts` file using the following commands from the root of your `server/` folder:

    ```
    mkdir src && touch src/index.ts
    ```

    Note that in the case of TypeScript, we're using the `.ts` file extension instead of the `.js` extension, which is used for plain JavaScript source files.

2. Next, open the `src/index.ts` file and add the following code, which is required to start a server on the `8080` port of localhost:

    ```
    import express, { Application } from 'express';
    const PORT = 8080;
    const app: Application = express();
    app.get('/', (req, res) =>
      res.send('Express is successfully running!'));
    app.listen(PORT, () => {
      console.log('Server is running at
                http://localhost:${PORT}');
    });
    ```

    In the preceding code, we start by importing `express` and `Application` from the `express` module and then define a constant to hold the port number where our server will be listening.

    Next, we create an application instance and add a route using the `get()` method of the Express application. This route will enable us to invoke a handler method when a GET request is made to the `/` endpoint. Finally, we call the `listen()` method of the Express application to start listening for HTTP requests on the provided port.

    We create routes in Express.js using some methods that are available on the Express application. These methods contain the names of the HTTP requests that can be handled by the route; for example, you can use `.get()` for GET requests, and `.post()` for POST requests. You can check the documentation at `https://expressjs.com/en/4x/api.html#app.METHOD` for a complete list.

    In the example we'll be building in this book, we don't need to use many routes because GraphQL only needs one endpoint that can listen to and handle all the requests in our app.

3. Now, we can run our server using `ts-node` by adding the following `start` script to our `package.json` file:

```
"scripts": {
  "start": "ts-node ./src/index.ts",
},
```

4. Head back to your Terminal and run the following command to invoke the previous script, which starts our server:

```
npm start
> [...]
Server is running at http://localhost:8080
```

If you visit the `http://localhost:8080` address with your web browser, you should see a blank page with the following message: **Express is successfully running!**

Now that we've created our Express.js server, we'll learn how to configure our project to watch for any code changes during development, automatically recompile our code, and restart the server.

Watching and recompiling our code

At this point, we can easily run our TypeScript code without compilation using `ts-node`. This is great for development, but every time we change our code, we'll need to stop and execute our script again.

This isn't appropriate when it comes to developing real-world projects, so we can take this a step further by using some utilities that watch the project's folder, such as `nodemon` and `ts-node-dev`. Let's use the latter because it's the easiest one to use.

Head back to your Terminal, stop your running server, and run the following command from your server application:

```
npm install --save-dev ts-node-dev
```

This command will install `ts-node-dev` as a development dependency.

Next, change the `start` script in the `package.json` file, as follows:

```
"scripts": {
  "start": "ts-node-dev --respawn ./src"
},
```

Next, run your `start` script again and change your `index.ts` file (you can simply add a comment, such as `// Our code is being watched and recompiled now thanks to ts-node-dev`). Your server should restart without doing this manually and it will output a message in the Terminal that is similar to the following:

```
[INFO] 16:57:48 Restarting: /home/ahmed/ngsocial/packages/
server/src/index.ts has been modified
Server is running at http://localhost:8080
```

This is a good setup for development, but when your project is ready for deployment, you only need to compile your TypeScript code to plain JavaScript once.

We can achieve this by adding another script that simply invokes the `tsc` compiler. Go back to the `package.json` file and add a script named `build` (you can choose any valid name you want), as follows:

```
"scripts": {
  "start": "ts-node-dev --respawn ./src",
  "build": "tsc --project ./"
}
```

The `--project` flag is used to specify the folder of the TypeScript files that will be compiled to JavaScript. Here, we specified `./` where the `tsconfig.json` file exists. This command will allow us to run the `tsc` compiler without specifying the full `./node_modules/typescript/bin/tsc` path of the compiler.

There are other ways to do this, such as installing TypeScript globally on our system using the `-g` switch of the `npm install` command or using `npx` to run `tsc` from the npm registry without installing it (`npx tsc`).

Head back to your Terminal, stop the running server, and let's invoke our `build` script using the following command from the root of your server project:

```
npm run build
```

This will run the `tsc` compiler to compile our project's files as plain JavaScript. At this point, we only have one file (`index.ts`), which will be converted into JavaScript code in the `dist/src/index.js` file.

We can have a flat output of the JavaScript source files inside the dist/ folder without the src/ subfolder by changing the value of the rootDir property in the tsconfig.json file to src/:

```
{
  "compilerOptions": {
    "target": "es6",
    "module": "commonjs",
    "rootDir": "./src",
    "outDir": "./dist",
    "esModuleInterop": true,
    "strict": true
  }
}
```

Now, if we run our npm run build command again, we'll get the output in the dist/ folder without the src/ subfolder.

This is the compiled code:

```
"use strict";
var __importDefault =
  (this && this.__importDefault) || function (mod) {
    return (mod && mod.__esModule) ? mod : { "default": mod };
};
Object.defineProperty(exports, "__esModule", { value: true
  });
const express_1 = __importDefault(require("express"));
const PORT = 8080;
const app = express_1.default();
app.get('/', (req, res) =>
  res.send('Express is successfully running!'));
app.listen(PORT, () => {
    console.log('Server is running at
                http://localhost:${PORT}');
});
```

We can test whether we can run our server with Node using the following command:

```
node dist/index.js
```

This should start the Express server and output **Server is running at http://localhost:8080** in the Terminal. If you visit that address on your browser, it will display a message stating **Express is successfully running!**.

From now on, we'll be working with the `ts-node-dev` utility (using the previous `npm start` command). This will allow us to continuously watch our TypeScript source code, which resides inside the `src/` folder of our server project and invokes the `ts-node` utility, to run it and re-run it after each change we make.

So far, we have initialized our server project, installed Express.js and TypeScript, and created a basic server running on the `8080` port of our localhost machine. In the next section, we'll learn how to configure Apollo Server to expose a GraphQL API.

Creating a GraphQL API

In this section, we'll learn how to create our GraphQL API using Apollo Server, an open source GraphQL server maintained by the community. First, we'll create a server with a simple schema. Then, we will add the types and resolvers to enable us to query and mutate posts, comments, and likes that are made by each user; we will use fake data first, before adding a real database.

Installing the necessary libraries

Let's start by installing the necessary libraries. Head back to your Terminal and run the following commands from the root of the `server/` folder:

```
npm install graphql apollo-server-express @graphql-tools/utils
@graphql-tools/schema
npm install --save-dev @types/graphql graphql-tag
```

The `graphql-tag` library exports a JavaScript template literal tag that converts GraphQL query strings to a **GraphQL Abstract Syntax Tree**.

The previous commands will install `graphql v15.6.0`, `apollo-server-express v3.3.0`, `@graphql-tools/schema v 8.2.0`, and `@graphql-tools/utils v8.2.3` as dependencies, plus `graphql-tag v2.12.5` and `@types/graphql v14.5.0` as development dependencies.

Next, let's learn how to proceed by creating our GraphQL server and schema. First, we'll expose a server with a simple API before adding the API for our social network application.

Exposing a simple GraphQL API

Now that we've installed the necessary libraries, we'll learn how to expose a simple GraphQL server with a simple schema to test how our server works:

1. Open the `src/index.ts` file and replace its contents with the following code as you progress.

   Start by adding the necessary imports:

   ```
   import express, { Application } from 'express';
   import { ApolloServer, Config, gql }
     from 'apollo-server-express';
   import { IResolvers } from '@graphql-tools/utils';
   ```

2. Next, add a type definition, as follows:

   ```
   const typeDefs = gql'
       type Query {
           message: String!
       }
     '
   ```

3. Add the resolver:

   ```
   const resolvers: IResolvers = {
     Query: {
       message: () => 'It works!'
     }
   };
   ```

4. Then, add the following configuration object:

   ```
   const config: Config = {
     typeDefs: typeDefs,
     resolvers: resolvers
   };
   ```

5. Finally, add the following function:

   ```
   async function startApolloServer(config: Config) {
     const PORT = 8080;
     const app: Application = express();
   ```

```
const server: ApolloServer =
    new ApolloServer(config);
await server.start();
server.applyMiddleware({
    app,
    path: '/graphql'
});
app.listen(PORT, () => {
    console.log('Server is running at
                http://localhost:${PORT}');
});
}
startApolloServer(config);
```

First, we start by adding the necessary imports, such as Express.js, from the express package, and `ApolloServer` and `gql` from the `apollo-server-express` packages. Next, we define our schema type and resolver and then create an `express` application.

After that, we create an instance of Apollo Server that takes `typeDefs` and `resolvers` as configuration parameters. Next, we add Apollo Server to the Express application by using `applyMiddleware`, which takes the `express` application and GraphQL endpoint as parameters.

You need to await `server.start()` before calling `server.applyMiddleware()` of `ApolloServer`.

Our schema contains one simple type definition that has a single field named `message` of the `string` type. In the query resolver object, we implemented a resolver function for the message field that simply returns some text stating **It works!**.

Important Note

This book assumes you are already familiar with the basic concepts of GraphQL. If that's not the case, or if you need a refresher, we recommend that you go to `https://graphql.org/learn/`.

Next, let's interact with this API by sending queries to our server using Apollo Studio.

Sending queries with Apollo Studio

Make sure your server is still running without any errors and then go to the `http://localhost:8080/graphql` address with your web browser. You should be presented with an interface that's similar to the following:

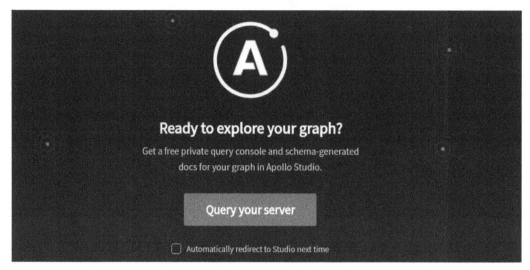

Figure 2.1 – Query your server

From this interface, you can click on the **Query your server** button to access Apollo Studio, where you can send requests to your API endpoint.

In the left panel, write the following GraphQL query to send a message query:

```
query {
  message
}
```

Click on the **Run** button. You should get the following output in the right-hand panel:

```
{
  "data": {
    "message": "It works!"
  }
}
```

The following screenshot shows Apollo Studio, along with our query and response:

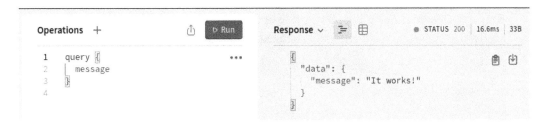

Figure 2.2 – Apollo Studio

Now that we've tested our Apollo Server, let's create the GraphQL schema of our social application. We'll be working with entities such as users, posts, comments, likes, and notifications.

Creating a GraphQL schema for our social network

In this section, we'll create the type definitions and resolvers in separate files and import them into our `index.ts` file. Let's change the folder structure of our previous example and make sure it still works as expected before adding more GraphQL types:

1. Head back to your Terminal and run the following commands from the root of your server folder:

    ```
    cd src && mkdir graphql
    cd graphql && touch schema.ts && touch schema.graphql &&
    touch resolvers.ts
    ```

2. Open the `src/graphql/schema.graphql` file and add the following type:

    ```
    type Query {
        message: String!
    }
    ```

3. Open the `src/graphql/resolvers.ts` file and add the following code:

    ```
    import { IResolvers } from '@graphql-tools/utils';
    const resolvers: IResolvers = {
      Query: {
        message: () => 'It works!'
      }
    };
    export default resolvers;
    ```

Here, we created a `resolvers` variable of the `IResolvers` type that contains the resolver for the `message` field and then exported it from the file.

4. Next, open the `src/graphql/schema.ts` file and add the following code:

```
import fs from 'fs';
import { GraphQLSchema } from 'graphql';
import { makeExecutableSchema }
  from '@graphql-tools/schema';
import { gql } from 'apollo-server-express';
import resolvers from './resolvers';

const typeDefs = gql'${fs.readFileSync(
  __dirname.concat('/schema.graphql'), 'utf8')}';
const schema: GraphQLSchema = makeExecutableSchema({
  typeDefs,
  resolvers,
});
export default schema;
```

Here, we read the query from the `.graphql` file using the `readFileSync` method imported from the `fs` module and imported the resolvers from the `resolvers.ts` file using the `import` statement. Next, we called the `makeExecutableSchema` with the `typeDefs` and `resolvers` arguments to create a schema that we can export from our `schema.ts` file.

> **Important Note**
> If you are using **Webpack**, you can import GraphQL queries from `.graphql` files using Webpack with a loader that comes with the `graphql-tag` package.

5. Open the `src/index.ts` file and update its contents, as follows:

```
import express, { Application } from 'express';
import { ApolloServer } from 'apollo-server-express';
import schema from './graphql/schema';
async function startApolloServer() {
  const PORT = 8080;
```

```
const app: Application = express();
const server : ApolloServer =
  new ApolloServer({schema});
await server.start();
server.applyMiddleware({
  app,
  path: '/graphql'
});
app.listen(PORT, () => {
  console.log('Server is running at
               http://localhost:${PORT}');
});
}
startApolloServer();
```

We import our schema using the import statement and pass it to our Apollo Server. If you run your server again, you should be able to interact with Apollo Studio, as you did previously.

Now, let's add the types for our small social network app.

6. Go to the src/graphql/schema.graphql file and add the User type, as follows:

```
type User {
  id: ID!
  fullName: String!
  bio: String
  email: String!
  username: String!
  password: String!
  image: String
  coverImage: String
  postsCount: Int!
  createdAt: String!
}
```

7. Next, add the `Post` type, as follows:

```
type Post {
    id: ID!
    text: String
    image: String
    author: User!
    commentsCount: Int!
    likesCount: Int!
    likedByAuthUser: Boolean
    latestComment: Comment
    latestLike: String
    createdAt: String!
}
```

8. Add the `Comment` type, as follows:

```
type Comment {
    id: ID!
    Comment: String!
    Author: User!
    post: ID!
    createdAt: String!
}
```

9. Add the `Like` type, as follows:

```
type Like {
    id: ID!
    user: User!
    post: ID!
    createdAt: String!
}
```

10. Add the `Notification` type, as follows:

```
type Notification {
    id: ID!
    text: String!
```

```
    postId: ID!
    createdAt: String!
}
```

11. Then, we need to add queries, as follows:

```
type Query {
  message: String!
  getUser(userId: ID!): User
  getPostsByUserId(userId: ID!, offset: Int, limit:
    Int): [Post]
  getFeed(offset: Int, limit: Int): [Post]
  getNotificationsByUserId(userId: ID!, offset: Int,
    limit: Int): [Notification]
  getCommentsByPostId(postId: ID!, offset: Int, limit:
    Int): [Comment]
  getLikesByPostId(postId: ID!, offset: Int, limit:
    Int): [Like]
  searchUsers(searchQuery: String): [User]
}
```

12. Finally, add the mutations, as follows:

```
type Mutation {
  post(text: String, image: String): Post
  removePost(id: ID!): ID
  comment(comment: String!, postId: ID!): Comment
  removeComment(id: ID!): Comment
  like(postId: ID!): Like
  removeLike(postId: ID!): Like
  removeNotification(id: ID!): ID
}
```

This is the initial schema that we'll improve upon as we progress in the next chapters. Before we continue implementing our GraphQL API, let's learn how to make use of mocking with Apollo Server (this is an important step if you plan to follow a **GraphQL-first development** process).

Mocking our GraphQL API

Mocking allows frontend developers to start building the UI without waiting for the backend development to finish. It also lets us test the UI without waiting on slow database operations or running a complete GraphQL server.

The community has provided many tools that make it easy to mock requests based on a GraphQL schema and queries such as `graphql-tools`, `graphql-faker`, and the built-in mocking feature of Apollo Server.

Since we are already using Apollo Server, let's use its mocking feature to mock our GraphQL requests before implementing the resolvers.

Open the `src/index.ts` file and add `mocks: true` to Apollo Server, as follows:

```
const server : ApolloServer = new ApolloServer({schema ,
    mocks: true});
```

The built-in mocking of Apollo Server is very basic. It simply looks at the `type` definitions and returns a value of the same type as the corresponding field.

Head over to Apollo Studio and add the following query:

```
query {
  searchUsers(searchQuery:"a") {
    id fullName email postsCount
  }
  getPostsByUserId(userId:"1") {
    id text author { id fullName } commentsCount
    latestComment { id comment }
    createdAt
  }
}
```

The following screenshot shows the query results:

```
Response ∨                                          ● STATUS 200 | 12.5ms | 823B

{
  "data": {
    "searchUsers": [
      {
        "id": "4f4555de-7e4d-4a59-90d7-5589f5321b86",
        "fullName": "Hello World",
        "email": "Hello World",
        "postsCount": 26
      },
      {
        "id": "62aeba1e-e5c3-4af0-8ce0-3e6267e33c87",
        "fullName": "Hello World",
        "email": "Hello World",
        "postsCount": -71
      }
    ],
    "getPostsByUserId": [
      {
        "id": "d854f2fb-7654-4d4a-aa0d-1bf00bd889f8",
        "text": "Hello World",
        "author": {
          "id": "18e61c97-d5eb-4b12-8744-49c570e2efe6",
          "fullName": "Hello World"
        },
        "commentsCount": 72,
        "latestComment": {
          "id": "2c081d3c-18d1-4c88-bcc6-3d787bf8e2bf",
          "comment": "Hello World"
        },
        "createdAt": "Hello World"
      }
```

Figure 2.3 – Query results with Apollo Studio

Here, we can see that, by default, each query, which is expected to return a list of values, returns only two values and that each field with a string value returns `Hello World`. This provides the right shape for the results but it's far from being realistic.

Fortunately, we can use other tools alongside Apollo Server, such as **casual**, a **fake data generator** for JavaScript that allows us to return more realistic and random values from GraphQL queries, even before implementing the required resolvers.

Head back to your Terminal and run the following command from the root of the `server/` folder:

```
npm install casual
```

Next, head back to the `src/index.ts` file and add the following import:

```
import casual from 'casual';
```

Next, add a `mocks` object, as follows:

```
const mocks = {};
```

Next, mock the User type by adding the following code inside the mocks object:

```
User: () => ({
    id: casual.uuid,
    fullName: casual.full_name,
    bio: casual.text,
    email: casual.email,
    username: casual.username,
    password: casual.password,
    image: 'https://picsum.photos/seed/picsum/200/300',
    coverImage:
        'https://picsum.photos/seed/picsum/200/300',
    postsCount: () => casual.integer(0)
}),
```

Next, mock the Post type, as follows:

```
Post: () => ({
    id: casual.uuid,
    text: casual.text,
    image: 'https://picsum.photos/seed/picsum/200/300',
    commentsCount: () => casual.integer(0),
    likesCount: () => casual.integer(0),
    latestLike: casual.first_name,
    createdAt: () => casual.date()
}),
```

Next, mock the Comment type, as follows:

```
Comment: () => ({
    id: casual.uuid,
    comment: casual.text,
    post: casual.uuid,
    createdAt: () => casual.date()
}),
```

Next, mock the `Like` type, as follows:

```
Like: () => ({
    id: casual.uuid,
    post: casual.uuid
})
```

For each field of each `type` in our schema, we can assign either a constant value or random value using `casual`, which provides generators for the most used types of data. You can find a list of the available properties and methods at `https://github.com/boolean/casual`.

By default, our mock resolvers return only two results for queries. We can change this default behavior as follows. In the same `mocks` object, you need to add the following object:

```
Query: () =>({
    getPostsByUserId: () =>
        [...new Array(casual.integer(10, 100))],
    getFeed: () => [...new Array(casual.integer(10, 100))],
    getNotificationsByUserId: () =>
        [...new Array(casual.integer(10, 100))],
    getCommentsByPostId: () =>
        [...new Array(casual.integer(10, 100))],
    getLikesByPostId: () =>
        [...new Array(casual.integer(10, 100))],
    searchUsers: () =>
        [...new Array(casual.integer(10, 100))]
})
```

Now, our queries will return between 10 and 100 fake entries. See `https://www.graphql-tools.com/docs/mocking#using-lists-in-mocks` for more details.

Update the object that was passed to `ApolloServer`, as follows:

```
const server: ApolloServer = new ApolloServer({schema,
    mocks, mockEntireSchema: false});
```

If you are working with a team of developers, the frontend developer can start building the frontend since they can interact with the GraphQL API and fetch the generated data without waiting for the backend developer to implement the resolvers. One issue that arises is that the entries that should be linked in a real-world scenario are not linked in our mocked API – let's learn how to change that!

Linking the mocked data

Posts should belong to users, so they should be linked to users via the user field. However, the randomly generated value of the field is not guaranteed to be one of the values that's generated for the users' identifiers. We can solve this easily by defining two arrays before the mocked object in the `src/index.ts` file, as follows:

```
let postsIds: string[] = [];
let sersIds: string[] = [];
```

These arrays will allow us to keep track of the generated identifiers for users and posts to ensure that the user and post fields of the `Post`, `Comment`, and `Like` types receive one of these values. This way, our generated data will be more realistic and similar to a real-world example.

Next, change the generator for the `id` field on the mock `User` object to the following function:

```
id: () => {let uuid = casual.uuid; usersIds.push(uuid);
        return uuid},
```

This makes sure that the generated ID is added to the `usersIds` array.

Next, you need to change the `id` field of the mocked `Post` type to the following function:

```
id: () => {let uuid = casual.uuid; postsIds.push(uuid);
        return uuid},
```

Similarly, this makes sure that the generated post ID is added to the `postsIds` array.

Next, add an `author` field to the mocked `Post` object, as follows:

```
author: casual.random_element(usersIds),
```

Next, change the `author` field of the mocked `Comment` object, as follows:

```
author: casual.random_element(usersIds),
```

Next, add a user field to the Like type, as follows:

```
user: casual.random_element(usersIds),
```

Next, change the post fields of the Comment and Like types, as follows:

```
post: casual.random_element(postsIds)
```

We've added all these modifications to make sure the fields that point to users and posts are getting values from the usersIds and postsIds arrays.

Finally, you need to change the user and author fields on the Post, Comment and Like types in our src/graphql/schema.graphql file to the ID type instead of the User type:

```
author: ID!
user: ID!
```

This is only temporary for the sake of mocking realistic data relationships.

Now, head back to Apollo Studio and run the following query:

```
query {
  searchUsers { id }
  getPostsByUserId(userId:"93c9c019"){ id author }
  getCommentsByPostId(postId:"93c9c019"){ id author post }
  getLikesByPostId(postId:"93c9c019"){ id user post }
}
```

Make sure that you pass the right values for userId and postId as they will be different in your case since they are randomly generated values.

You should get linked results – that is, the user and post fields will contain values that are equal to the values that were generated for the posts' identifiers.

Now that we have realistic and linked data to test our API, we'll learn how to configure CORS. This will help us connect to our GraphQL server from the Angular frontend without getting blocked by the browser.

Configuring CORS

Cross-origin resource sharing (**CORS**) refers to a mechanism by which we can allow or disallow other domains from requesting restricted resources on a web page that are being served from a specific domain.

We have already set up a server on port 8080 of our localhost machine and we'll be also running a development server of the Angular CLI from another port in the next few chapters. Different ports on localhost are considered different origins, so we need to configure CORS in our Express.js server; otherwise, the requests from our Angular frontend will be disallowed.

Fortunately for us, configuring CORS in Express.js is easy. Head to your Terminal and run the following command from the root of the server folder to install the cors module:

```
npm install cors
```

Now, you need to open the src/index.ts file and import the cors module:

```
import cors from 'cors';
```

Next, add the following line just below where you created an instance of the Express.js application:

```
app.use(cors());
```

This is the basic configuration of CORS in our Express.js server, which allows any other domains to request data from your server without restriction.

You can refer to the docs of the cors module at https://expressjs.com/en/resources/middleware/cors.html for more options for configuring CORS while protecting your server resources from non-authorized access.

Summary

In this chapter, we started implementing our backend application, which will be exposing the GraphQL API. We created a project using Lerna to follow the monorepo approach to organizing our code base, which will contain both the server and client apps.

We installed Express.js and configured it with TypeScript and also GraphQL via Apollo Server.

Next, we created an initial GraphQL schema for our small social network application and configured Apollo Server with mocking. Then, we used the casual tool to use realistic values for our mock data.

Thanks to mocking, we were able to run a working GraphQL server by providing only a schema without implementing the resolvers.

This is a great feature that allows us to easily test the GraphQL API without running a complete API server. It also allows the frontend developer/team to start working on the features that require data to be fetched from the server as quickly as possible without waiting for the backend development to finish.

Next, we used Apollo Studio to send queries to our Apollo Server, which was able to return data that contains the shapes of our schema types.

Finally, we installed the `cors` module and configured CORS in our Express/Apollo Server to enable our frontend application, which will be running from a different port at localhost during development, to send requests without getting blocked due to the same-origin policy in web browsers.

In the next chapter, we'll continue our journey of building our backend application to expose a fully working GraphQL API with resolvers that fetches data from a real MySQL database.

3
Connecting the Database with TypeORM

In the last chapter, we learned how to set up a Node.js server with GraphQL support to implement a backend application. Additionally, we explained how to install Express.js and configure it with TypeScript and GraphQL. We used mocking to provide a working GraphQL server with Apollo Server before implementing the resolvers that should actually be responsible for fetching and adding data.

In addition, we set up Cross-Origin Resource Sharing (CORS) and used Apollo Studio to communicate with our GraphQL server.

In this chapter, we'll look at how to connect a MySQL database to our application using TypeORM, as well as how to create resolvers to get and save data from the database.

Now that we've built a backend application that can respond to GraphQL queries, we'll need to connect it to our database in order to store our data. We'll utilize TypeORM to abstract the database operations, allowing you to use any desired database management system for your app without changing the code. In addition, we will learn how to connect TypeORM with Apollo, as well as how to generate and seed our database tables.

We will discuss the following topics in this chapter:

- Creating a MySQL user and database
- Setting up TypeORM and MySQL in our server project
- Creating TypeORM entities
- Seeding test data in our database
- Using Apollo together with TypeORM

Technical requirements

Before you can proceed with the steps in this chapter, you must first finish the steps in the previous chapter on how to set up a functional GraphQL server. Please refer to *Chapter 2, Setting Up GraphQL with Node.js, Express.js, and Apollo*, to learn how to build up a GraphQL server using Express.js and Apollo Server.

It is important to note that your work machine should have MySQL installed. Please refer to *Chapter 1, App Architecture and Development Environment*, for instructions.

You should also be familiar with the following technologies:

- TypeScript
- Node.js and npm
- SQL (including basic concepts such as tables, columns and their types, relationships, and primary and foreign keys)
- GraphQL basics (including resolvers, queries, and mutations)

If you want to learn more about SQL, please see `https://www.khanacademy.org/computing/computer-programming/sql`.

You should also go through TypeORM's official documentation, which is accessible at `https://typeorm.io/#/`.

If you need a review on the fundamentals of GraphQL, please see `https://graphql.org/learn/`.

The code for this chapter may be found in the GitHub repository at `https://github.com/PacktPublishing/Full-Stack-App-Development-with-Angular-and-GraphQL/tree/main/Chapter03`.

Creating a MySQL user and database

When we first installed MySQL on our development machine, it created a root user who can manage the database system. This user has complete access to the MySQL server, including all databases and tables.

Furthermore, beginning with MySQL v5.7, the root user can be authenticated via the auth socket plugin rather than a username and password, as allowed by the MySQL client used with TypeORM.

> **Important Note**
> The name of the system user that launches the MySQL client and the MySQL user must be the same when using the auth socket plugin; moreover, you must execute mysql with sudo privileges to have access to your database(s).

For the reasons stated above, it is required to create a new user in MySQL and provide it with the right permissions. To do so, go through the following steps:

1. To begin, use the following command to connect to MySQL:

   ```
   sudo mysql
   ```

   You'll be asked to enter a password. Enter your system's root password.

2. Once you've accessed the MySQL prompt, you may use SQL commands such as CREATE USER to create a new user with a password that utilizes the mysql native password plugin for authentication:

   ```
   CREATE USER 'dbuser'@'localhost' IDENTIFIED WITH mysql_
   native_password BY 'p4ssw0rd';
   ```

   In this case, we created a user called dbuser with the password p4ssw0rd.

3. We must provide it with the necessary permissions using the following command:

   ```
   GRANT ALL PRIVILEGES ON * . * TO dbuser@'localhost';
   ```

   We have given the user all of the privileges (that is, full root access to everything in our database). This is not advised for security reasons. However, keep in mind that this is simply a test database.

4. Then, use the following command to reload all of the privileges:

   ```
   FLUSH PRIVILEGES;
   ```

5. Exit the current MySQL prompt:

```
exit;
```

6. Use the following command to log in as the new user:

```
mysql -u dbuser -p
```

Enter the password given to the user, in other words, p4ssw0rd.

7. Finally, to create a database, use the following command:

```
create database socialdb;
```

That's all! We have a user who goes by the name dbuser, with the p4ssw0rd password and a socialdb database. These credentials will be used to set up TypeORM in the following section.

Setting up TypeORM and MySQL

In this section, we'll install and configure TypeORM in our server project.

TypeORM is a Node.js and TypeScript **Object Relational Mapper** (**ORM**). It lets you build and query database tables with TypeScript and high-level programming structures such as classes, decorators, and functions, rather than SQL commands.

Return to your Terminal. To start the server, make sure you're within the server's project and then perform the following commands:

```
cd ngsocial/packages/server/
npm start
```

Ascertain that you are inside the server's project. After that, execute the following commands in another Terminal:

```
cd ngsocial/packages/server/
npm install typeorm reflect-metadata mysql
```

The reflect-metadata library is essential in providing support for the TypeScript decorators required by TypeORM.

Next, open the tsconfig.json file and make the following changes:

```
{
  "compilerOptions": {
```

```
    "target": "es6",
    "module": "commonjs",
    "rootDir": "./src",
    "outDir": "./dist",
    "esModuleInterop": true,
    "strict": true,
    "experimentalDecorators": true,
    "emitDecoratorMetadata": true,
    "strictPropertyInitialization": false
  }
}
```

We added the experimentalDecorators and emitDecoratorMetadata attributes and set them to true. We also added strictPropertyInitialization and set it to false.

Because decorators are currently an experimental feature, the experimentalDecorators attribute provides experimental support for them. The emitDecoratorMetadata attribute tells the compiler to emit the metadata necessary for decorated declarations. To produce decorator metadata, we must import the reflect-metadata package into our TypeScript code.

The strictPropertyInitialization attribute instructs the compiler whether strict property initialization checks should be enabled or disabled in classes.

Then, open the src/index.ts file and begin modifying the imports, as shown here:

```
import express, { Application } from 'express';
import { ApolloServer } from 'apollo-server-express';
import schema from './graphql/schema';
import cors from 'cors';
import 'reflect-metadata';
import { createConnection, Connection } from 'typeorm';
```

Then, edit the rest of the code as follows:

```
const connection: Promise<Connection> = createConnection();
connection.then(() => {
  startApolloServer();
}).catch(error => console.log("Database connection error:
        ", error));
```

```
async function startApolloServer() {
  const PORT = 8080;
  const app: Application = express();
  app.use(cors());
  const server: ApolloServer = new ApolloServer({ schema
  });
  await server.start();
  server.applyMiddleware({
    app,
    path: '/graphql'
  });
  app.listen(PORT, () => {
    console.log('Server is running at
      http://localhost:${PORT}');
  });
}
```

To provide support for the decorators, we imported the `reflect-metadata` library. Following that, we imported the `createConnection` function from the TypeORM package and deleted several symbols, such as casual, which is no longer required in this part.

The `reflect-metadata` package is a utility library that enhances the TypeScript decorators' capabilities. This is essential for TypeORM to function correctly.

The `mocks` object was then discarded, and the line where the Apollo Server is created was changed to remove the `mocks` parameter. We don't need mocks in this part because we'll be creating actual resolvers.

We used the `createConnection` function to create a TypeORM connection and then invoked the `then()` method of the returned promise. Next, in the body of the promise's `then()` method, we invoked the code to build the Express and Apollo Server. This ensures that the server is only started after a successful connection to the MySQL database.

If a problem occurs, a database connection error message containing the error details will be printed to the console.

Because we have already launched our server in a previous step, we'll get the Database connection error: TypeORMError: No connection options were found in any orm configuration files error after saving the file.

To resolve this issue, use the following command to create the ormconfig.json file in the server/ folder:

```
touch ormconfig.json
```

Then, in the ormconfig.json file, add the following contents:

```
{
    "type": "mysql",
    "host": "localhost",
    "port": 3306,
    "username": "dbuser",
    "password": "p4ssw0rd",
    "database": "socialdb",
    "synchronize": true,
    "logging": false,
    "entities": [
        "src/entity/**/*.ts"
    ],
    "migrations": [
        "src/migration/**/*.ts"
    ],
    "subscribers": [
      "src/subscriber/**/*.ts"
    ]
}
```

This is the TypeORM configuration file, where we add the database connection configuration settings, such as the database type (in our instance, mysql), the database host and post, the database name, the username, and the password.

After that, if everything is properly configured, stop and restart the development server using the following command:

```
npm start
```

So, we've now configured TypeORM. Following that, we'll go over how to create the database entities for our social application.

Creating TypeORM entities

In this part, we'll create entities that represent our social network database. An entity is a TypeScript class that has been annotated with the @Entity() decorator that corresponds to a database table.

When working with a database using plain SQL rather than an ORM, you begin by creating database tables. With TypeORM, we must also begin by creating entities.

Add the @Entity() decorator to the class definition and the @Column (or equivalent) decorator to each property in the class to create an entity. These annotations provide TypeORM with the information it needs to create the database table and columns.

We can also use the following decorators to establish associations between the entities:

- OneToOne: This is used to create a one-to-one relationship between two objects.
- JoinColumn: This specifies which part of the relationship contains the join column.
- OneToMany: This specifies a one-to-many relationship between objects.
- ManyToOne: This specifies a many-to-one relationship between objects.
- ManyToMany: This is used to establish a many-to-many relationship among objects.
- JoinTable: This specifies the join columns of a many-to-many relationship.

Create a subdirectory named entity/ in the src/ folder by running the following command:

```
mkdir src/entity
```

We can simply think about our database structure based on the schema structure because we already defined our GraphQL schema types based on our app requirements.

As a result, the following entities will be necessary in our case:

- User
- Post
- Comment

- Like
- Notification

To define an entity, we must first create a TypeScript file under the `entity/` folder. A basic entity is made up of one or more columns, one of which should be the primary column.

Begin by creating an `src/entity/User.ts` file and then adding the following entity:

```
import { Entity, Column, PrimaryGeneratedColumn,
CreateDateColumn } from 'typeorm';

@Entity()
export class User {
  @PrimaryGeneratedColumn() id: number;
  @Column() fullName: string;
  @Column("text", { nullable: true }) bio: string;
  @Column({ unique: true }) email: string;
  @Column({ unique: true }) username: string;
  @Column() password: string;
  @Column({ nullable: true }) image: string;
  @Column({ nullable: true }) coverImage: string;
  @Column({ default: 0 }) postsCount: number;
  @CreateDateColumn() createdAt: Date;
}
```

As previously stated, `@Entity()` instructs TypeORM to match the class to a database table. To make it easier for TypeORM to discover the entity, we place it in the `entity/` folder. This was previously configured in the `ormconfig.json` file's `entities` array.

To produce the matching database's table columns, we just use the `@Column` decorator to decorate the class property that you wish to make into a column.

TypeORM will map a string to a varchar-like type (depending on the database type) and a number to an integer-like type by default. We may also modify the default type by supplying the required type to the column decorator. It's worth noting that we utilized the text type for the `bio` column rather than the limited varchar type.

TypeORM will build database tables with the same names as the entities; however, if you want to use a different table name, you may give it in the `@Entity` decorator — for example, `@Entity("mytablename")`.

The `@PrimaryGeneratedColumn()` decorator defines a primary `Int` column that is automatically generated with an auto-incrementing value.

We added the `default` option to the `postsCount` column, which specifies that it has a default value of zero; the `nullable` option to the `bio`, `image`, and `coverImage` columns, which specifies that they can be null; and the `unique` option to the `username` and `email` columns, which specifies that duplicate values are not accepted.

The `@CreateDateColumn()` decorator adds the entity's insertion date to the `createdAt` column automatically.

The `id` column, which includes the primary key of the corresponding database table, and the `createdAt` column, which contains the date of insertion, will be shared by all entities.

For building the next entities, we'll use the same decorators and parameters.

For representing a user's post, create an `src/entity/Post.ts` file and add the following entity:

```
import { Entity, Column, PrimaryGeneratedColumn,
CreateDateColumn } from 'typeorm';

@Entity()
export class Post {
  @PrimaryGeneratedColumn() id: number;
  @Column("longtext") text: string;
  @Column({ nullable: true }) image: string;
  @Column({ default: 0 }) commentsCount: number;
  @Column({ default: 0 }) likesCount: number;
  @Column({ default: "" }) latestLike: string;
  @CreateDateColumn() createdAt: Date;
  @Column({ default: false }) likedByAuthUser: boolean;
}
```

In this case, we added the `longtext` option to the `text` column in order to modify the column type from the default (`varchar(255)` for string fields) to `longtext`.

Next, create the `src/entity/Comment.ts` file and add the following entity to model a user's comment:

```
import { Entity, Column, PrimaryGeneratedColumn,
CreateDateColumn } from 'typeorm';
```

```
@Entity()
export class Comment {
  @PrimaryGeneratedColumn() id: number;
  @Column("text") comment: string;
  @CreateDateColumn() createdAt: Date;
}
```

To represent a user's like, create an `src/entity/Like.ts` file and add the following entity:

```
import { Entity, PrimaryGeneratedColumn, CreateDateColumn }
from 'typeorm';

@Entity('likes')
export class Like {
  @PrimaryGeneratedColumn() id: number;
  @CreateDateColumn() createdAt: Date;
}
```

In the `Entity` decorator, we used `likes` for the table name instead of the name of `like` which is a reserved keyword in SQL.

Finally, add the following entity to the `src/entity/Notification.ts` file to model a user's notification:

```
import { Entity, Column, PrimaryGeneratedColumn,
CreateDateColumn } from 'typeorm';

@Entity()
export class Notification {
  @PrimaryGeneratedColumn() id: number;
  @Column() text: string;
  @Column() postId: number;
  @CreateDateColumn() createdAt: Date;
}
```

The code for this part may be found in the GitHub commit at `https://git.io/JzddQ`.

Following that, we must build certain relationships between the entities in accordance with the following requirements:

- A user may have one or more notifications, posts, comments, and likes.

- A post may be associated with one or more comments and likes, an author, and the most recent comment.

Return to the `src/entity/User.ts` file and include the following imports:

```
import { OneToMany } from 'typeorm';
import { Post } from './Post';
import { Comment } from './Comment';
import { Like } from './Like';
import { Notification } from './Notification';
```

We just imported the entities that will be associated with the user entity and the decorator that will be used to construct the one-to-many relationship in this case.

Add the following fields next:

```
@Entity()
export class User {
  // [...]
  @OneToMany(type => Post, post => post.author) posts:
    Post[];
  @OneToMany(type => Comment, comment => comment.author)
    comments: Comment[];
  @OneToMany(type => Like, like => like.user) likes:
    Like[];
  @OneToMany(type => Notification, notification =>
    notification.user) notifications: Notification[];
}
```

Here, we added some extra fields of the array type, including posts, comments, likes, and notifications with `@OneToMany` annotations. This is one of the provided entity relationships by TypeORM.

TypeORM inserts the necessary foreign keys between tables when using entity relationships.

We modeled the requirement where a user may have many comments, likes, and notifications.

In addition, we utilized the @OneToMany decorator to construct a one-to-many relationship between two entities.

The type => Post statement is a function that returns the associated entity's class. Because of language constraints, this approach is used instead of utilizing the class directly. () => Post is another option.

Return to the src/entity/Post.ts file and insert the following imports:

```
import { OneToOne, OneToMany, ManyToOne, JoinColumn } from
'typeorm';
import { User } from './User';
import { Comment } from './Comment';
import { Like } from './Like';
```

We imported the associated user, comment, and like entities, as well as the decorators, in order to create one-to-one, one-to-many, and many-to-one relations.

Add the following fields next:

```
@Entity()
export class Post {
  // [...]
  @OneToOne(type => Comment, comment => comment.post, {
    onDelete: 'SET NULL' })
  @JoinColumn() latestComment: Comment;
  @ManyToOne(type => User, user => user.posts, { onDelete:
    'CASCADE' }) author: User;
  @OneToMany(type => Comment, comment => comment.post)
    comments: Comment[];
  @OneToMany(type => Like, like => like.post) likes:
    Like[];
}
```

We added extra fields for author, latest comment, comments, and likes with @OneToMany, @ManyToOne, and @OneToOne annotations.

We modeled the requirement where a post might have multiple comments, multiple likes, one most recent comment, and one author.

To define which end holds the relationship's column, we utilized the `@JoinColumn` decorator. It's the post table in this example. This is essential and should be present on just one end of the relation.

Return to the `src/entity/Comment.ts` file and insert the following imports:

```
import { ManyToOne } from 'typeorm';
import { User } from './User';
import { Post } from './Post';
```

Here, we imported the user and post entities with which we want to form the associations, as well as the many-to-one decorator.

Add the following fields next:

```
@Entity()
export class Comment {
  // [...]
  @ManyToOne(type => User, user => user.comments, {
    onDelete: 'CASCADE' })
  author: User;
  @ManyToOne(type => Post, post => post.comments, {
    onDelete: 'CASCADE' })
  post: Post;
}
```

We added author and post fields with @ManyToOne annotations here. This implies that a single person could add many comments, and a post can have several comments. These are the inverse relationships specified for the user and post entities.

Return to the `src/entity/Like.ts` file and include the following imports:

```
import { ManyToOne } from 'typeorm';
import { User } from './User';
import { Post } from './Post';
```

To define the inverse relationships, we imported the user and post entities, as well as the many-to-one decorator.

Add the following fields next:

```
@Entity('likes')
export class Like {
  // [...]
  @ManyToOne(type => User, user => user.likes, { onDelete:
    'CASCADE' })
  user: User;
  @ManyToOne(type => Post, post => post.likes, { onDelete:
    'CASCADE' })
  post: Post;
}
```

Note that we also added extra fields for a user and a post with @ManyToOne annotations to define the inverse relationships that we previously created for the user and post entities.

Return to the src/entity/Notification.ts file and include the following imports:

```
import { ManyToOne } from 'typeorm';
import { User } from './User';
```

Here, we imported the user entity with which we want to construct the inverse relation, as well as the many-to-one decorator.

Then, include the following field:

```
@Entity()
export class Notification {
  // [...]
  @ManyToOne(type => User, user => user.notifications, {
    onDelete: 'CASCADE' })
  user: User;
}
```

With a @ManyToOne annotation, we added the field for the user to whom the notification belongs. This is the inverse of the relationship defined in the user entity.

In this entity, we modeled the requirement where one or more notifications may belong to a user.

The code for adding TypeORM relations can be found in the GitHub commit at https://git.io/JzdFl.

So we've built the entities required to establish our social application database, as well as the associations between them. Next, we'll go over how to populate the database tables with test data, which will allow us to easily test the functionality of our resolvers in the next sections.

Seeding test data

So far, we've built database entities that let TypeORM construct SQL tables. Let's now enter some information into the database. You could do this manually, but it would be exhausting; we need to automate it sufficiently so that we can have it up and running quickly.

We can accomplish this by utilizing the typeorm-seeding package to create factories and seeders for our entities, which simplifies the process. Perform the following steps:

1. Return to your Terminal and begin by installing the package with the following command:

    ```
    npm install typeorm-seeding
    ```

2. Install the type definitions of the Faker library and create the src/database/, database/seeds/, and database/factories/ folders:

    ```
    npm install -D @types/faker
    cd src/ && mkdir database
    cd database && mkdir seeds factories
    ```

3. Add the following scripts to the package.json file to invoke the seed commands:

    ```
    "scripts": {
      "seed:config": "ts-node ./node_modules/typeorm-
        seeding/dist/cli.js config",
      "seed:run": "ts-node ./node_modules/typeorm-
        seeding/dist/cli.js seed",
      [...]
    }
    ```

4. Add the following commands to the package.json file to make it easier for you to drop and sync the database:

    ```
    "scripts": {
      "schema:drop": "ts-node
    ```

```
    ./node_modules/typeorm/cli.js schema:drop",
  "schema:sync": "ts-node
    ./node_modules/typeorm/cli.js schema:sync",
  [...]
}
```

To create and populate the database with test data, a seeder class is utilized. It has a single method named run where we may enter data into our database using TypeORM's query builder or entity factory.

To produce the data, we can utilize entity factories instead of individually supplying the properties for each entity seed.

For each entity, we need to create a factory where you can create the entity and return it. Those factory files should be created in the src/database/factories folder and suffixed with .factory.ts:

1. Create the factory for the user entity. Create an src/database/factories/ user.factory.ts file and add the following code:

```
import Faker from 'faker';
import { define } from 'typeorm-seeding';
import { User } from '../../entity/User';

define(User, (faker: typeof Faker) => {
  const user = new User()
  user.fullName = faker.name.findName();
  user.bio = faker.lorem.sentences();
  user.email = faker.internet.email();
  user.username = faker.internet.userName();
  user.password = faker.internet.password();
  user.image = faker.image.imageUrl();
  user.coverImage = faker.image.imageUrl();
  user.postsCount = 200;
  user.createdAt = faker.date.past();
  return user;
});
```

2. Create a factory for the post entity in the `src/database/factories/post.factory.ts` file:

```
import Faker from 'faker';
import { define, factory } from 'typeorm-seeding';
import { Post } from '../../entity/Post';

define(Post, (faker: typeof Faker) => {
  const post = new Post();
  post.text = faker.lorem.text();
  post.image = faker.image.imageUrl();
  post.commentsCount = 100;
  post.likesCount = 200;
  post.latestLike = faker.name.findName();
  post.createdAt = faker.date.past();
  return post;
});
```

3. Create the comment factory in the `src/database/factories/comment.factory.ts` file using the following code:

```
import Faker from 'faker';
import { define, factory } from 'typeorm-seeding';
import { Comment } from '../../entity/Comment';

define(Comment, (faker: typeof Faker) => {
  const comment = new Comment();
  comment.comment = faker.lorem.text();
  comment.createdAt = faker.date.past();
  return comment;
});
```

4. Create the like factory in the `src/database/factories/like.factory.ts` file, as follows:

```
import Faker from 'faker';
import { define } from 'typeorm-seeding';
import { Like } from '../../entity/Like';
```

```
define(Like, (faker: typeof Faker) => {
  const like = new Like();
  like.createdAt = faker.date.past();
  return like;
});
```

5. Create the notification factory in the `src/database/factories/`
    `notification.factory.ts` file, as follows:

```
import Faker from 'faker';
import { define } from 'typeorm-seeding';
import { Notification } from '../../entity/Notification';

define(Notification, (faker: typeof Faker) => {
  const notification = new Notification();
  notification.text = faker.lorem.words();
  notification.postId = 1;
  notification.createdAt = faker.date.past();
  return notification;
});
```

Following that, you must create a seeder class in the `src/database/seeds/`
`create-users.seed.ts` file:

6. Begin by including the following imports:

```
import { Seeder, Factory } from 'typeorm-seeding';
import { User } from '../../entity/User';
import { Post } from '../../entity/Post';
import { Comment } from '../../entity/Comment';
import { Like } from '../../entity/Like';
import { Notification } from '../../entity/Notification';
```

7. Define the seeder class, as follows:

```
export default class CreateUsers implements Seeder {
  public async run(factory: Factory): Promise<void> {
  }
}
```

8. Inside the `run` method, add the following code to create `15` users using the `createMany` method of the user factory:

```
await factory(User)().map(async (user: User) => {
    return user;
}).createMany(15);
```

9. Inside the `map` method, add the following code to create a random number of comments between 1 and 10:

```
const comments: Comment[] = await factory
    (Comment)().map(async (comment: Comment) =>
        {
            comment.author = await
                factory(User)().create();
            return comment;
        }).createMany(Math.floor(Math.random() * 10)
    + 1);
```

10. Similarly, create a random number of likes between 1 and 10:

```
const likes: Like[] = await
    factory(Like)().map(async (like: Like) => {
        like.user = await
            factory(User)().create();
        return like;
    }).createMany(Math.floor(Math.random() * 10)
    + 1);
```

11. Following this, create a random number of posts between 1 and 10, and associate the previously created comments and likes to these posts using the following code:

```
const userPosts: Post[] = await
    factory(Post)().map(async (post: Post) => {
        post.comments = comments;
        post.likes = likes;
        post.latestComment = await
            factory(Comment)().map(async (comment:
        Comment) => {
            comment.author = await
```

```
                factory(User)().create();
            return comment;
        }).create();
        return post;
    }).createMany(Math.floor(Math.random() * 10)
    + 1);
```

12. Create notifications and associate them with the posts and the current user, as follows:

```
const postIds = userPosts.map((post: Post)
    => post.id);
const notifications: Notification[] = await
    factory(Notification)().map(async
    (notification: Notification) => {
        const postId: number = postIds.pop() as
            number;
        notification.postId = postId;
        return notification;
    }).createMany(postIds.length);
```

13. Finally, associate the posts and the notifications with the current user in the map body, as follows:

```
user.posts = userPosts;
user.notifications = notifications;
return user;
```

That's all! This class will be in charge of producing all of the necessary fake data and associating the entries among them, the same as in a real-world scenario.

Following that, we'll need to execute the following command to start the seeder:

`npm run seed:run`

The code for this section may be found in the commit at https://git.io/JzdbJ. For more information, check out https://www.npmjs.com/package/typeorm-seeding.

Let's implement the resolvers now that we've filled the database with some test data.

Using Apollo with TypeORM

After we've created the entities and inserted some test data, we'll look at how to integrate Apollo with TypeORM. In this section, we'll define our GraphQL resolvers, which are functions that will be called when we send queries and mutations to our GraphQL endpoint:

1. First, you need to change back the `user` and `author` fields on the `Post`, `Comment`, and `Like` types of our GraphQL schema, in the `src/graphql/ schema.graphql` file, to the type of `User` instead of `ID`:

    ```
    author: User!
    user: User!
    ```

2. Additionally, change the type of the `post` field from `ID` to `Post` in the `Comment` and `Like` types:

    ```
    post: Post!
    ```

Before implementing our schema resolvers, we need to create the types for ensuring the type safety of our resolvers' code. Fortunately, we don't have to do this manually. Instead, we can use **GraphQL Code Generator**, which is a tool for generating code from your GraphQL schema and operations with a simple CLI. For additional information about this tool, please refer to `https://www.graphql-code-generator.com/`.

Since those types are generated code, we'll place them in their own package and then import them from our resolvers.

Following that, we'll look at how to use Lerna to create a package and configure it to be able to import the generated types:

1. If your server is still running, stop it and execute the following commands to create a `graphql-types` package:

    ```
    cd ../../
    lerna create graphql-types
    ```

    Note that you'll be asked some questions regarding your package. Accept the default values and press *Enter*.

    A new `graphql-types` package will be created in `./packages/graphql-types`.

Now that we have multiple packages, which are folders with a `package.json` file, in our project, let's use npm workspaces, which are supported by npm v7+, to efficiently manage dependencies across all of our monorepo project's packages.

2. Simply add the `workspaces` key to the `ngsocial/package.json` root file to inform npm where our packages exist:

```
{
    "name": "@ngsocial/ngsocial",
    "private": true,
    "devDependencies": {
        "lerna": "^4.0.0"
    },
    "version": "0.0.1",
    "workspaces": ["./packages/*"]
}
```

Observe that we also changed the name of the root package to `@ngsocial/ngsocial` to create a scoped package. For more information, please refer to `https://docs.npmjs.com/cli/v7/using-npm/scope`.

Lerna and npm can now both find packages in the `./packages` folder. If many packages require the same dependency, thanks to npm workspaces, it may be installed once in the root and imported from the other packages without trouble.

Continuing with our steps:

1. Install GraphQL Codegen and the required plugins in the root of our monorepo project, as shown here:

```
npm install @graphql-codegen/cli @graphql-codegen/
typescript @graphql-codegen/typescript-resolvers
```

It is worth noting that we may install these dependencies within the `graphql-types` package as well; however, the official documentation advises installing Codegen within the root. For further details, please see `https://www.graphql-code-generator.com/docs/getting-started/development-workflow`.

2. Create an ngsocial/codegen.yml file and add the following contents:

```
schema: './packages/server/src/graphql/schema.graphql'
generates:
  ./packages/graphql-types/src/resolvers-types.ts:
    plugins:
      - typescript
      - typescript-resolvers
```

3. Open the ngsocial/package.json file and add the following script:

```
"scripts": {
    "codegen": "npx graphql-codegen --watch"
},
```

This script will launch GraphQL Codegen in watching mode.

Then, use the following command to start it:

npm run codegen

This will create a graphql-types/src/resolvers-types.ts file with all of the required types for our schema resolvers.

Because this is a TypeScript file contained within another package, we must be able to import it into our server's code, which is contained within a different package. To do this, we may use a tsconfig.json file with the appropriate settings and an index.ts file.

4. Create a packages/graphql-types/tsconfig.json file and add the following contents:

```
{
    "compilerOptions": {
      "outDir": "lib",
      "rootDir": "src",
      "baseUrl": "./",
      "composite": true
    },
    "include": ["src/**/*"],
    "exclude": ["node_modules", "**/__tests__/*"]
}
```

The `"composite": true` option is required to be able to reference this package from the other packages.

Next, create a `packages/graphql-types/src/index.ts` file and add the following code:

```
export * from './resolvers-types';
```

This ensures that we export all of the symbols that we need.

Next, you need to open the `graphql-types/package.json` file and change name to `@ngsocial/graphql` to create a scoped package.

5. Finally, you need to open the `server/tsconfig.json` file and update it, as follows:

```
{
    "compilerOptions": {
      [...]
      "paths": {
        "@ngsocial/graphql": ["../graphql-types/src"]
      }
    },
    "references": [{ "path": "../graphql-types" }]
}
```

You need to ensure that the pathname specified in the `paths` object matches the name in the `graphql-types/package.json` file, which, in our case, is `@ngsocial/graphql`.

That's all! We can now import our generated GraphQL types by writing the following:

```
import { Resolvers, User, QueryGetUserArgs } from '@ngsocial/graphql';
```

To ensure consistent naming throughout our project's packages, edit the `server/package.json` file and change the name to `@ngsocial/server`.

You can see the updates of these steps on GitHub at `https://git.io/Jzd77`.

Following that, let's perform a few more steps:

1. Create a `server/src/entity/index.ts` file and add the following imports:

    ```
    export { User } from './User';
    export { Post } from './Post';
    export { Comment } from './Comment';
    export { Like } from './Like';
    export { Notification } from './Notification';
    ```

    Instead of importing each entity from its associated file, we simply added a barrel file that allows us to import our entities with short paths, as seen here:

    ```
    import { User, Post, Comment, Like, Notification } from
    './entity';
    ```

    Please refer to `https://basarat.gitbook.io/typescript/main-1/barrel` for additional information about barrel files.

2. Open the `server/src/index.ts` file and add the highlighted imports:

    ```
    import { createConnection, Connection, Repository,
    getRepository } from 'typeorm';
    import { User, Post, Comment, Like, Notification } from
    './entity';
    ```

    We'll use the `Repository` and `getRepository()` symbols from the `typeorm` package to work with our entities.

    In TypeORM, a repository is an abstraction that provides a set of operations to query and save entities to the database. For more information, please check out `https://typeorm.io/#/working-with-repository` and `https://typeorm.io/#/entities`.

    Each entity has a repository that contains the methods for working with the entity. The repository can be created using the `getRepository()` method and then passing the entity, as follows:

    ```
    userRepository = getRepository(User);
    ```

3. Define and export the `Context` type, as follows:

    ```
    export type Context = {
      orm: {
        userRepository: Repository<User>;
        postRepository: Repository<Post>;
    ```

```
    commentRepository: Repository<Comment>;
    likeRepository: Repository<Like>;
    notificationRepository: Repository<Notification>;
  };
};
```

This type will be used to type the context object that will be passed to the resolvers. At this point, we only pass an `orm` object containing the TypeORM repositories for working with the database.

4. Add the following code in the `startApolloServer()` method right underneath the `app.use(cors());` line:

```
const userRepository: Repository<User> =
  getRepository(User);
const postRepository: Repository<Post> =
  getRepository(Post);
const commentRepository: Repository<Comment> =
  getRepository(Comment);
const likeRepository: Repository<Like> =
  getRepository(Like);
const notificationRepository:
  Repository<Notification> =
  getRepository(Notification);
```

Next, define the `context` object of the `Context` type and add the TypeORM repositories to it, as follows:

```
const context: Context = {
  orm: {
    userRepository: userRepository,
    postRepository: postRepository,
    commentRepository: commentRepository,
    likeRepository: likeRepository,
    notificationRepository: notificationRepository
  }
};
```

5. Finally, as shown in the following code, replace the code line where we create Apollo Server by providing the context with the schema:

```
const server: ApolloServer = new ApolloServer({ schema,
context });
```

To work with the database, we can now use this context object in the resolvers together with the provided repositories. We could alternatively call the `getRepository()` function for each entity in the resolvers, but this is the preferred solution for a variety of reasons, including easy testability. You can view the changes of the previous steps from the commit at `https://git.io/JzdQM`.

Following that, we'll implement the resolvers for our GraphQL schema:

1. Open the `server/src/graphql/resolvers.ts` file and update the imports, as follows:

```
import { Context } from '..';
import {
  Resolvers,
  User,
  Post,
  Comment,
  Like,
  Notification } from '@ngsocial/graphql';
import { ApolloError } from 'apollo-server-errors';
```

2. Update the type of the `resolvers` object to `Resolvers` instead of `IResolvers`:

```
const resolvers: Resolvers = {
  Query: {
    message: () => 'It works!'
  }
};
export default resolvers;
```

All of our schema resolvers and their arguments will be type safe as a result of this.

We utilize our custom `Context` type for the context object of the resolvers because it is typed with `any`. We can also tell GraphQL Codegen to utilize this type for the context. Please refer to `https://the-guild.dev/blog/better-type-safety-for-resolvers-with-graphql-codegen` for additional details.

Next, inside the `Query` object, add the resolver functions listed below, one by one, separated by a comma:

1. Begin by including the `getUser` query resolver, which is used to find a user by ID:

```
getUser: async (_, args, ctx: Context) => {
  const orm = ctx.orm;
  const user = await orm
    .userRepository
    .findOne({
      where:
        { id: args.userId }});
  if (!user) {
    throw new ApolloError(
      "No user found",
      "USER_NOT_FOUND");
  }
  return user as unknown as User;
}
```

To find the user with the specified ID, we utilize the `findOne()` method of the user repository, which is obtained from the `orm` object passed with the resolver's context. If no user is found, an Apollo error is thrown.

We cast the user object from the `User` entity to the `User` type before returning it, and because TypeScript warns that the types do not sufficiently overlap, we first cast the object to the `unknown` type.

2. Implement the `getPostsByUserId` resolver to retrieve posts based on the user ID:

```
getPostsByUserId: async (_, args, ctx: Context) => {
  const posts = await ctx.orm.postRepository
    .createQueryBuilder("post")
    .where({ author: { id: args.userId } })
    .leftJoinAndSelect("post.author", "post_author")
```

```
      .leftJoinAndSelect("post.latestComment",
        "latestComment")
      .leftJoinAndSelect("latestComment.author",
        "latestComment_author")
      .leftJoinAndSelect("post.likes", "likes")
      .leftJoinAndSelect("likes.user", "likes_user")
      .orderBy("post.createdAt", "DESC")
      .skip(args.offset as number)
      .take(args.limit as number)
      .getMany();
    return posts as unknown as Post[];
  }
```

The query builder in TypeORM allows you to create and execute basic and advanced SQL queries. In this case, post is an alias for selected posts. Aliases are used to get access to the selected data's columns.

Please refer to https://typeorm.io/#/select-query-builder for more information about how to work with the query builder.

3. Implement the getFeed resolver to obtain a feed of all network users' posts:

```
getFeed: async (_, args, ctx: Context) => {
  const feed = await ctx.orm.postRepository
    .createQueryBuilder("post")
    .leftJoinAndSelect("post.author", "post_author")
    .leftJoinAndSelect("post.latestComment",
      "latestComment")
    .leftJoinAndSelect("latestComment.author",
      "latestComment_author")
    .leftJoinAndSelect("post.likes", "likes")
    .leftJoinAndSelect("likes.user", "likes_user")
    .orderBy("post.createdAt", "DESC")
    .skip(args.offset as number)
    .take(args.limit as number)
    .getMany();
  return feed as unknown as Post[];
}
```

4. Implement the `getNotificationsByUserId` resolver, which is used to retrieve notifications belonging to a user based on their ID:

```
getNotificationsByUserId: async (_, args, ctx: Context)
=> {
  const notifications = await
    ctx.orm.notificationRepository
      .createQueryBuilder("notification")
      .innerJoinAndSelect("notification.user", "user")
      .where("user.id = :userId", { userId: args.userId
      })
      .orderBy("notification.createdAt", "DESC")
      .skip(args.offset as number)
      .take(args.limit as number)
      .getMany();
  return notifications as unknown as Notification[];
}
```

5. Implement the `getCommentsByPostId` resolver to obtain comments for a post based on its ID:

```
getCommentsByPostId: async (_, args, ctx: Context) => {
  return await ctx.orm.commentRepository
    .createQueryBuilder("comment")
    .innerJoinAndSelect("comment.author", "author")
    .innerJoinAndSelect("comment.post", "post")
    .where("post.id = :id",
      { id: args.postId as string })
    .orderBy("comment.createdAt", "DESC")
    .skip(args.offset as number)
    .take(args.limit as number)
    .getMany() as unknown as Comment[];
}
```

6. Implement the `getLikesByPostId` resolver to obtain likes for a post based on its ID:

```
getLikesByPostId: async (_, args, ctx: Context) => {
  return await ctx.orm.likeRepository
    .createQueryBuilder("like")
    .innerJoinAndSelect("like.user", "user")
    .innerJoinAndSelect("like.post", "post")
    .where("post.id = :id", { id: args.postId })
    .orderBy("like.createdAt", "DESC")
    .skip(args.offset as number)
    .take(args.limit as number)
    .getMany() as unknown as Like[];
}
```

7. Implement the `searchUsers` resolver to search for users by their names or usernames:

```
searchUsers: async (_, args, ctx: Context) => {
  const users = await ctx.orm.userRepository
    .createQueryBuilder("user")
    .where('user.fullName Like
      '%${args.searchQuery}%'')
    .orWhere('user.username Like
      '%${args.searchQuery}%'')
    .getMany();
  return users as unknown as User[];
}
```

Now that we've implemented the queries, let's move on to the mutations. Add the following methods to the `Mutation` object:

1. Begin by including the following resolvers:

```
post: (_, args, ctx: Context) => {
  throw new ApolloError("Not implemented yet",
    "NOT_IMPLEMENTED_YET");
},
comment: (_, args, ctx: Context) => {
  throw new ApolloError("Not implemented yet",
```

```
    "NOT_IMPLEMENTED_YET");
},
like: (_, args, ctx: Context) => {
  throw new ApolloError("Not implemented yet",
    "NOT_IMPLEMENTED_YET");
},
removeLike: async (_, args, ctx: Context) => {
  throw new ApolloError("Not implemented yet",
    "NOT_IMPLEMENTED_YET");
}
```

We just throw an Apollo error here, indicating that the resolver has not yet been implemented in this chapter. We'll implement them in the following chapter, once we've implemented authentication, because we need information about the current (authenticated) user.

2. Implement the removePost resolver to delete a post from the database by its ID, as seen here:

```
removePost: async (_, args, {orm}: Context) => {
  const post = await orm.postRepository.findOne(
    args.id, { relations: ['author'] });
  if(!post){
    throw new ApolloError("Post not found",
      "POST_NOT_FOUND");
  }
  const result: DeleteResult = await
    orm.postRepository.createQueryBuilder()
    .delete().from(PostEntity).where("id = :id", { id:
      args.id }).execute();
  const postsCount = post?.author?.postsCount;
  if (postsCount && postsCount >= 1) {
    await orm.userRepository.update({ id:
      post?.author.id }, { postsCount: postsCount - 1
      });
  }
  if(result.affected && result.affected <= 0) {
    throw new ApolloError("Post not deleted",
```

```
    "POST_NOT_DELETED");
  }
  return args.id;
}
```

For additional details, please check out `https://typeorm.io/#/delete-query-builder`.

3. Implement the `removeComment` resolver to delete a comment by its ID from the database:

```
removeComment: async (_, args, {orm}: Context) => {
  const comment = await orm.commentRepository.findOne(
    args.id, {relations:['author', 'post']});
  if(!comment){
    throw new ApolloError("Comment not found",
      "COMMENT_NOT_FOUND");
  }
  const result: DeleteResult = await
    orm.commentRepository.delete(args.id);
  if(result.affected && result.affected <= 0) {
    throw new ApolloError("Comment not deleted",
      "COMMENT_NOT_DELETED");
  }
  const commentsCount = comment?.post?.commentsCount;
  if(commentsCount && commentsCount >= 1){
    await orm.postRepository.update(comment.post.id, {
      commentsCount: commentsCount - 1 });
  }
  return comment as unknown as Comment;
}
```

4. Implement the `removeNotification` resolver to delete a notification by its ID from the database:

```
removeNotification: async (_, args, {orm}: Context) => {
  const notificationRepository =
    orm.notificationRepository;
  const notification = await
```

```
    notificationRepository.findOne(args.id,
    {relations:['user']});
  if(!notification){
    throw new ApolloError("Notification not found",
      "NOTIFICATION_NOT_FOUND");
  }
  const result: DeleteResult = await
    notificationRepository.delete(args.id);
  if(result.affected && result.affected <= 0) {
    throw new ApolloError("Notification not deleted",
      "NOTIFICATION_NOT_DELETED");
  }
  return args.id;
}
```

It's worth noting that we didn't include a mutation to create notifications. This is due to the fact that they must be created whenever people comment or like content. As a result, we can simply create them whenever new comment or like entities are added.

We can do this by using the `AfterInsert` decorator on some of the `Comment` and `Like` entities' methods, which will be called after the entities are inserted into the database:

1. Open the `server/src/entity/Comment.ts` file and add the following imports:

    ```
    import { AfterInsert, getRepository } from 'typeorm';
    import { Notification } from './Notification';
    ```

2. Add the `createNotification()` method to the class and annotate it with the `AfterInsert` decorator, as shown:

    ```
    @AfterInsert()
    async createNotification() {
      if (this.post && this.post.id) {
        const notificationRepository =
          getRepository(Notification);
        const notification =
          notificationRepository.create();
    ```

```
notification.user = await getRepository(
  User).createQueryBuilder("user")
  .innerJoinAndSelect("user.posts", "post")
  .where("post.id = :id", { id: this.post.id })
  .getOne() as User;
notification.postId = this.post.id;
notification.text = '${this.author.fullName}
  commented on your post';
await notificationRepository.save(notification);
  }
}
```

Using the methods of the notification repository, we create the necessary notification within the method.

Then we must repeat the process on the Like entity:

1. Add the following imports to the server/src/entity/Like.ts file:

    ```
    import {  AfterInsert, getRepository } from 'typeorm';
    import { Notification } from './Notification';
    ```

2. Add an @AfterInsert-decorated method where you must implement the code to provide a suitable notification informing users that someone liked their post:

    ```
    @AfterInsert()
    async createNotification() {
      if (this.post && this.post.id) {
        const notificationRepository =
          getRepository(Notification);
        const notification =
          notificationRepository.create();
        notification.user = await getRepository(
          User).createQueryBuilder("user")
          .innerJoinAndSelect("user.posts", "post")
          .where("post.id = :id", { id: this.post?.id })
          .getOne() as User;
        notification.postId = this.post?.id;
        notification.text = '${this.user.fullName} liked
          your post';
    ```

```
      await notificationRepository.save(notification);
    }
  }
```

Using the repository's `create()` method, we built a `Notification` object and filled up the necessary properties such as the user, post ID, and text. Furthermore, we utilized TypeORM's query builder to obtain the user who should receive the notification, which is simply the person who created the post.

You may play around with the resolvers that we've built using Apollo Studio. If your server is already up and running, go to `https://studio.apollographql.com/sandbox/explorer` to start sending queries and mutations:

1. If you send one of the `post`, `comment`, `like`, or `removeLike` mutations, you should get an error message with the Not implemented yet text and the `NOT_IMPLEMENTED_YET` code:

```
mutation {
  post(text: "Hello World"){
    id
  }
  comment(comment: "This is a comment", postId: 100 )
  {
    id
  }
  like(postId: 100) {
    id
  }
  removeLike(postId: 100) {
    id
  }
}
```

To send a mutation, you need to use the `mutation` operation type and then specify one mutation or multiple mutations to send. For more information, please refer to `https://graphql.org/learn/queries/`.

2. You can send any of the following queries individually or all at once. Begin by sending the get Feed query to retrieve a list of posts published on the network:

```
query {
  getFeed(offset: 0 limit: 10) {
    id text commentsCount likesCount author { id
      fullName }
  }
}
```

We must use the query operation type in this case. We use the schema's get Feed field to request a feed of posts from the server. We pass in the offset and limit variables to instruct the server to deliver only a subset of the available posts, rather than all of them and we utilize a selection set to request the subfields that we require from the server. To obtain the required fields from the nested author object, we also must use a selection set.

3. You can then take a user ID and pass it to the following query to obtain the user's posts. In my case:

```
getPostsByUserId(userId:224 limit: 5) {
  id text author { id fullName } createdAt
}
```

4. To access the user's information, such as their full name, bio, and post count, run the following query:

```
getUser(userId: 224) {
  id fullName postsCount bio
}
```

5. You may obtain notifications for a specific user by running the following query:

```
getNotificationsByUserId(userId: 19) {
  id text createdAt
}
```

6. Using the post ID and the following queries, you may retrieve the likes and comments on a certain post:

```
getCommentsByPostId(postId: 10 limit: 5) {
  id comment author { id fullName } post { id }
```

```
    }
    getLikesByPostId(postId: 17 limit: 5) {
      id user { id fullName } post { id }
    }
```

7. Finally, you may delete posts, comments, and notifications by executing the following mutations:

```
mutation {
  removePost(id: 59)
  removeComment(id: 21) {
    id
  }
  removeNotification(id: 10)
}
```

In this case, we send mutations to delete the post with ID 59, the comment with ID 21, and the notification with ID 10.

We must use a selection set for the removeComment mutation since the mutation returns the removed comment object. The other two mutations return the deleted post or notification's ID (a scalar).

The response we received was as follows:

```
{
  "data": {
    "removePost": "59",
    "removeComment": {
      "id": "21"
    },
    "removeNotification": "10"
  }
}
```

This indicates that we were successful in removing these objects.

Summary

In this chapter, we learned how to use TypeORM to connect a MySQL database to our application, as well as how to implement some resolvers to get and remove data from the database. This enabled us to write a functional GraphQL API with resolvers that get data from a real MySQL database.

We used TypeORM to abstract database operations, which allows you to utilize any chosen database management system for your application without changing the code.

We looked at how to generate and populate our database tables with initial data before integrating TypeORM with Apollo.

This chapter is now complete! In the following chapter, we'll continue building our backend application with Apollo Server and Node.js to implement authentication and image uploading.

4
Implementing Authentication and Image Uploads with Apollo Server

In the previous chapter, we covered how to connect a MySQL database to our web application using TypeORM and we implemented some of the resolvers for communicating with the database. We've been able to expose a working GraphQL API with actual resolvers that query and remove data from a real MySQL database.

We used TypeORM for working with the database instead of plain SQL, which abstracts the underlying database management system and SQL instructions using high-level programming constructs in TypeScript.

We have created the TypeORM entities responsible for generating the database tables and columns for our social application and integrated TypeORM with Apollo Server. In this chapter, we'll see how to add authentication and image uploads with Apollo Server to our GraphQL API and we'll implement more resolvers.

We'll learn about the necessary concepts for adding authentication with Node.js, Express, and Apollo Server and then how to handle image uploads.

We will specifically cover the following topics:

- What's JWT?
- Implementing authentication with Node.js and Apollo
- Protecting GraphQL queries and mutations
- Implementing image uploading

Technical requirements

To complete the steps of this chapter, you are required to have completed *Chapter 3, Connecting the Database with TypeORM*, of this book.

You need to be familiar with the following technologies:

- TypeScript
- Node.js
- Basics of SQL
- Basics of GraphQL

You need to have an account with Amazon S3/DigitalOcean Spaces Object Storage or any compatible service.

Amazon provides a free tier that you can use to follow the steps of this chapter, but if you are unable to sign up for an account with them, you can use tools such as MinIO to run a server locally that's compatible with Amazon S3. Check out `https://docs.min.io/` for more information.

You can find the source code of this chapter from at `https://github.com/PacktPublishing/Full-Stack-App-Development-with-Angular-and-GraphQL/tree/main/Chapter04`.

What's JWT?

JWT stands for **JSON Web Token**, and it's an open standard (RFC 7519) that offers a concise and self-contained mechanism for securely encoding and transmitting information between computers as JSON objects.

When developing GraphQL APIs, you'll almost always need to secure specific endpoints from public access and need users to be authenticated first and authorized (allowed) to use the endpoint(s).

You can achieve that using JWTs. This is how authentication with JWTs works:

1. When users register or sign in, we build a JWT token that includes the user's identity on the server and send it to the client.

2. We need to transmit the received token back to the server with every request on the client side so that the server can verify the client's identity and determine whether they are permitted to visit the endpoint.

In contemporary web apps, where we have a JavaScript client app that has to interface with an API to query and save data to the server, JWT authentication is quite popular and efficient.

Because tokens are not encrypted but merely signed using cryptographic techniques, using HTTPS with JWT is needed when persisting the user's data to secure it from any man-in-the-middle attacks that may intercept and manipulate it.

Authentication may be accomplished in a number of ways. The most typical method is to ask for the user's email and password using a login form that the user fills out.

When protecting our APIs, we should note the difference between two concepts:

*   **Authentication**, which is the process of confirming a user's identity.

*   **Authorization**, which is used to verify whether the authenticated user is authorized to access or perform certain actions based on their role on the system.

Now that we understand what a JWT token is and the difference between authentication and authorization, let's see how to implement authentication with Apollo.

Implementing authentication

You'll almost always need to safeguard queries and mutations from unauthenticated and/or unauthorized users while implementing your GraphQL API. In this section, we'll add authentication with JWT to our GraphQL API.

In Apollo, we can retrieve the JWT token sent back by the client from the HTTP header, extract the user information from that token, and include it in the context, which can be accessed from any resolver. In the resolver, we can use the user's information to verify what data the user is authorized to access.

We'll be using libraries such as `dotenv` for loading environment variables from a `.env` file that we can use to add the secret required for generating JWTs, and `jsonwebtoken` for generating the tokens. We'll also be using the Scrypt algorithm for hashing the users' passwords before saving them to the database.

Let's now see the practical steps:

1. Let's get started by installing the required libraries using the following commands from the folder of your server project:

    ```
    npm install dotenv jsonwebtoken
    npm install --save-dev @types/jsonwebtoken
    ```

2. Add the following type to your GraphQL schema, available from the `server/src/graphql/schema.graphql` file:

    ```
    type AuthPayload {
        token: String!
        user: User!
    }
    ```

3. Next, we need to add two mutations for registering and signing in users to the `src/graphql/schema.graphql` file:

    ```
    type Mutation {
      [...]
      register(fullName: String!, username: String!,
          email: String!, password: String!): AuthPayload!
      signIn(email: String!, password: String!):
          AuthPayload!
    }
    ```

4. After updating the schema, you need to regenerate the types. Open a new command-line interface and run the following command from the folder of your monorepo project (where the `/packages` folder lives):

    ```
    npm run codegen
    ```

5. After that, we need to implement the resolvers in the `server/src/graphql/resolvers.ts` file. Start by adding the following imports:

```
import jsonwebtoken from 'jsonwebtoken';
import crypto from 'crypto';
import dotenv from 'dotenv';
```

6. Next, add the following line:

```
dotenv.config();
```

This will configure `dotenv` to load our environment variable(s).

7. Inside the `server/` folder, create a `.env` file and add the following variable that will be used to sign the tokens:

```
JWT_SECRET=azerty123456
```

We used `azerty123456` for our JWT shared secret. However, for production, make sure to use a strong key for signing JWTs with the HMAC algorithm to prevent any brute-force attacks.

8. Add the following method for hashing passwords in the `server/src/graphql/resolvers.ts` file:

```
const hashPassword = async (plainPassword: string):
Promise<string> => {
  return new Promise((resolve, reject) => {
    const bytes = crypto.randomBytes(16);
    const salt = bytes.toString("hex");
    crypto.scrypt(plainPassword, salt, 64,
    (error, buffer) => {
      if (error) reject(error);
      const hashedPassword =
      '${salt}:${buffer.toString('hex')}';
      resolve(hashedPassword);
    });
  })
};
```

We'll use this function to hash our passwords before saving them to the database using Scrypt – a salted hashing algorithm.

In order to hash passwords using Scrypt, you need to create a unique salt for each hash. You can check out `https://auth0.com/blog/adding-salt-to-hashing-a-better-way-to-store-passwords/` and `https://qvault.io/cryptography/very-basic-intro-to-the-scrypt-hash/` for more details.

This function concatenates and returns both the salt and password's hash, which will both be saved to the database because the salt is required to verify the password.

9. Add the following method, which will be used to compare the plain password to the hash:

```
const compareToHash = async (plainPassword: string, hash:
string): Promise<boolean> => {
  return new Promise((resolve, reject) => {
    const result = hash.split(":");
    const salt = result[0];
    const hPass = result[1];
    crypto.scrypt(plainPassword, salt, 64,
      (error, buffer) => {
        if (error) reject(error);
        resolve(hPass == buffer.toString('hex'));
      });
  })
};
```

Check out the docs `https://nodejs.org/api/crypto.html#crypto_crypto_scrypt_password_salt_keylen_options_callback`.

10. Next, add the `register` mutation for registering users:

```
register: async (_, args, { orm }) => {}
```

Inside the body of the method, we first destructure the `args` object to retrieve the user's full name, username, email, and password:

```
const { fullName, username, email, password } = args;
```

11. Next, we call the `create()` method of the user's repository to create a new user with the retrieved information. For the password, we call the `hashPassword()` method to hash it before assigning it:

```
let user = orm.userRepository.create({
  fullName: fullName,
```

```
    username: username,
    email: email,
    password: await hashPassword(password),
    postsCount: 0,
    image: 'https://i.imgur.com/nzTFnsM.png'
});
```

12. Next, we call the `save()` method of the user's repository to save the user in the database:

```
const savedUser = await orm.userRepository.save(user)
.catch((error: unknown) => {
  if (error instanceof QueryFailedError &&
    error.driverError.code == "ER_DUP_ENTRY") {
    throw new ApolloError("A user with this
      email/username already exists",
      "USER_ALREADY_EXISTS");
  }
});
```

Make sure you import the `QueryFailedError` class, as follows:

```
import { DeleteResult, QueryFailedError } from 'typeorm';
```

After persisting the registered user information, we create a JWT from the user's ID and email with a one-year expiration period and a secret key, using the `sign()` method of the `jsonwebtoken` library:

```
const token = jsonwebtoken.sign(
  { id: savedUser.id, email: user.email },
  process.env.JWT_SECRET as string,
  { expiresIn: '1y' }
);
```

For more details about JWTs, read `https://jwt.io/introduction/`.

13. Finally, we return the generated JWT and the user from the resolver:

```
return { token: token, user: savedUser };
```

14. Next, let's implement the `signIn` mutation, as follows:

```
signIn: async (_, args, { orm }) => {}
```

Inside the method's body, we destructure the args object to get the user's email and password:

```
const { email, password } = args;
```

Then, call the findOne() method of the user's repository to verify whether a user exists in the database with the provided email. If the user doesn't exist, we throw an error with the No user found with this email! message:

```
const user = await orm.userRepository.findOne({ where: {
email: email } });
if (!user) {
   throw new ApolloError('No user found with this
      email!', 'NO_USER_FOUND');
}
```

15. Next, we call the compareToHash() function to verify that the submitted password matches the returned user's password. If they don't match, we throw an error with the Incorrect password! message:

```
if (!await compareToHash(password, user.password)) {
   throw new ApolloError('Incorrect password!',
      'INCORRECT_PASSWORD');
}
```

16. After that, we generate a JWT token from the user's information and return both the user and token from the resolver:

```
const token = jsonwebtoken.sign(
   { id: user.id, email: user.email },
   process.env.JWT_SECRET as string,
   { expiresIn: '1d' }
);
return { token, user };
```

Next, visit http://localhost:8080/graphql and call your register and signIn mutations from Apollo Studio to see whether they are working as expected.

17. Start with the `register` mutation to create a user with the provided information:

```
mutation {
    register(
        fullName: "Ahmed Bouchefra"
        username: "ahmed.bouchefra"
        email: "ahmed@email.com"
        password: "123456789"
    ) {
        token
        user {
            id fullName username email password
        }
    }
}
```

This should create a user in the database and return the token, plus the ID, full name, and password of the saved user. The `fullName`, `username`, and `email` values submitted via the mutation parameters should be the same as the fields of the user's sub-selection, while the `password` value must be returned in an encrypted format.

For example, this is the response I received for the previous mutation:

Figure 4.1 – The response of the register mutation

If you rerun the mutation again with the same input, you should get an error stating that a user with that email or username already exists.

18. Next, send the `signIn` mutation to authenticate the user:

```
mutation {
  signIn(email: "ahmed@email.com", password:
    "123456789") {
    token
    user {
      id email password
    }
  }
}
```

This should verify that the user exists in the database with the submitted information and return it with a token. You should receive a token with user information. The email should be the same in the parameter and response, while the returned password should be hashed.

If you provide a wrong email or password that doesn't match the user's registered password, you should get the respective `No user found with this email!` or `Incorrect password!` error messages.

This is an example of the response we receive:

Figure 4.2 – The response of the signIn mutation

We have created the mutations for registering and signing in users using JWTs. Next, we'll see how to use the generated JWTs with Apollo to protect the GraphQL endpoints.

Protecting GraphQL queries and mutations

After implementing the GraphQL mutations for registering and signing users, we need to protect our other queries and mutations using the JWT token that we obtain when users are registered or signed in.

Following user authentication, the server will transmit a JWT token to the client, which should store it locally and send it back with each request using an HTTP header (the authorization header).

We can use the context option of Apollo Server to access the authorization header from the request object and pass this data to the resolvers. The context is available in each resolver.

The following are the steps required to protect our GraphQL API:

1. Open the `server/src/index.ts` file and add the following code:

    ```
    import jwt from 'jsonwebtoken'
    import dotenv from 'dotenv';
    dotenv.config();
    const { JWT_SECRET } = process.env;
    ```

    We import the necessary symbols and retrieve the `JWT_SECRET` environment variable from the `process.env` object (loaded during runtime from the `.env` file using `dotenv`). This will enable us to get our JWT secret from the `.env` file. Check out `https://nodejs.dev/learn/how-to-read-environment-variables-from-nodejs` for more details about how to read environment variables in Node.js.

2. Add the following method, which will be used to verify and retrieve the user's information from the JWT using the `verify()` method, from `jsonwebtoken`, and the secret:

    ```
    const getAuthUser = (token: string) => {
      try {
        if (token) {
          return jwt.verify(token, JWT_SECRET as string);
        }
        return null;
      } catch (error) {
        return null;
    ```

```
    }
  }
```

Next, we need to confirm the identity of the user before allowing them to access the protected endpoints.

3. Update the `Context` type, as follows:

```
export type Context = {
  orm: {
    userRepository: Repository<User>;
    postRepository: Repository<Post>;
    commentRepository: Repository<Comment>;
    likeRepository: Repository<Like>;
    notificationRepository: Repository<Notification>;
  };
  authUser: User | null;
};
```

4. Modify the code for bootstrapping Apollo Server, as follows:

```
const server: ApolloServer = new ApolloServer({ schema,
context: ({req}) => {
  const token = req.get('Authorization') || '';
  const authUser = getAuthUser(token.split(' ')[1]);
  const ctx: Context = {
    orm: {
      userRepository: userRepository,
      postRepository: postRepository,
      commentRepository: commentRepository,
      likeRepository: likeRepository,
      notificationRepository: notificationRepository
    },
    authUser: authUser
  };
  return ctx;
}});
```

In the context function, we call the get() method of the request object to retrieve the value of the authorization header, which should contain the "JWT:" string, plus the token. Next, we split the value using space and pass the second part, which contains the actual JWT, to our getAuthUser() function to decode the token and get the user's information, that is, the user's ID and email. Then, we create a ctx object (typed with Context) with the authUser and the orm object that contains the TypeORM repositories.

Please note that we've moved the creation of the context object from outside to inside the context method of Apollo Server, so you must remove the previous code for creating the instance that exists right above the call for bootstrapping Apollo Server. Check this GitHub commit for the updates we've made: https://git.io/JgCgO.

After implementing this, if a JWT is sent back with any request to our GraphQL API, we'll be able to access the logged-in user's ID and email from the context parameter of any resolver function.

5. After that, you can go to the server/src/graphql/resolvers.ts file and modify every resolver that you need to protect by adding the following:

```
getUser: async (_, args, { orm, authUser }) => {
    if(!authUser) throw new Error('Not authenticated!');
    // [...]
}
```

We destructure the context parameter passed to the function to get the authUser object that contains information about the authenticated user. Then, we check whether that object is not null; otherwise, we throw an error with the Not authenticated! message.

This will ensure that every resolver will be protected against unauthenticated users.

A better solution is using directives, so let's see how to implement this second approach. We'll implement an @auth directive that can be applied to the query and mutation fields in the schema to check whether the user is authenticated before they can complete the operation:

1. Open the server/src/graphql/schema.ts file and import the following symbols:

```
import { defaultFieldResolver } from 'graphql';
import { getDirective, MapperKind, mapSchema } from '@
```

```
graphql-tools/utils';
import { ApolloError } from 'apollo-server-express';
```

2. Next, add the following function:

```
function authDirective() {
  return {
    authDirectiveTypeDefs: gql'directive @auth on
      FIELD_DEFINITION',
    authDirectiveTransformer: (schema: GraphQLSchema)
      => mapSchema(schema, {})
  };
}
```

3. Next, add the following code to the second parameter of the mapSchema()
    function, between the curly braces:

```
[MapperKind.OBJECT_FIELD]: (fieldConfig) => {
  const directiveName = 'auth';
  const authDirective = getDirective(schema,
    fieldConfig, directiveName)?.[0];
  if (authDirective) {
    const { resolve = defaultFieldResolver } =
      fieldConfig;
    fieldConfig.resolve = (source, args, context,
      info) => {
      if (!context.authUser) {
        throw new ApolloError("User is not
          authenticated", "USER_NOT_AUTHENTICATED");
      }
      return resolve(source, args, context, info);
    };
    return fieldConfig;
  }
}
```

For more details about implementing directives, read https://www.graphql-
tools.com/docs/schema-directives.

4. Next, call the function and destructure the returned object, as follows:

```
const { authDirectiveTypeDefs, authDirectiveTransformer }
= authDirective();
```

We get the type definition of the authorization directive and the transformer function that will be applied to the schema to use the @auth directive.

5. Next, update the call to the makeExecutableSchema() function and include the type definition of the authorization directive, as follows:

```
const typeDefs = gql'${fs.readFileSync(__dirname.
concat('/schema.graphql'), 'utf8')}';
let schema: GraphQLSchema = makeExecutableSchema({
  typeDefs: [authDirectiveTypeDefs, typeDefs],
  resolvers: resolvers
});
```

Here, we pass an array to the typeDefs option that contains both the type definitions for our schema and the authorization directive.

6. Next, apply authDirectiveTransformer() to the schema before exporting it:

```
schema = authDirectiveTransformer(schema);
export default schema;
```

Check out the corresponding GitHub commit at https://git.io/JgCwA.

7. Finally, you need to apply the created @auth directive to the schema fields you need to protect. Open the server/src/graphql/schema.graphql file and update the Query type, as follows:

```
type Query {
  message: String!
  getUser(userId: ID!): User
  getPostsByUserId(userId: ID!, offset: Int, limit:
    Int): [Post] @auth
  getFeed(offset: Int, limit: Int): [Post] @auth
  getNotificationsByUserId(userId: ID!, offset: Int,
    limit: Int): [Notification] @auth
  getCommentsByPostId(postId: ID!, offset: Int, limit:
    Int): [Comment] @auth
  getLikesByPostId(postId: ID!, offset: Int, limit:
```

```
      Int): [Like] @auth
    searchUsers(searchQuery: String): [User] @auth
}
```

8. Next, update the Mutation type, as follows:

```
type Mutation {
    post(text: String, image: String): Post @auth
    removePost(id: ID!): ID @auth
    comment(comment: String!, postId: ID!): Comment
      @auth
    removeComment(id: ID!): Comment @auth
    like(postId: ID!): Like @auth
    removeLike(postId: ID!): Like @auth
    removeNotification(id: ID!): ID @auth
    register(fullName: String!, username: String!,
      email: String!, password: String!): AuthPayload!
    signIn(email: String!, password: String!):
      AuthPayload!
}
```

You can test that your implementation is working by sending a JWT with the **Authorization** header, which you can get after registering for a new account or signing in, using the **Headers** panel in Apollo Studio:

Figure 4.3 – Sending a JWT in Apollo Studio

For implementing a more complex permission system, take a look at *graphql-shield*, available from https://github.com/maticzav/graphql-shield.

For the post, comment, like, and removeLike mutations, we don't only need to check whether the user is authenticated but also use the authenticated user's ID to perform the action.

So, we need to add and destructure the context parameter to get the authenticated user's information.

9. Go back to the `server/src/graphql/resolvers.ts` file and add the following imports:

```
import {
    Post as PostEntity,
    Comment as CommentEntity,
    Like as LikeEntity } from '../entity';
```

10. Implement the `post` mutation, as follows:

```
post: async (_, args, { orm, authUser }: Context) => {
  const post = orm.postRepository.create(
    {
      text: args.text,
      image: args.image,
      author: await orm.userRepository.findOne(
        authUser?.id)
    } as unknown as PostEntity
  );
  const savedPost = await
    orm.postRepository.save(post);
  await orm.userRepository.update({ id: authUser?.id
    }, { postsCount: post.author.postsCount + 1 });
  return savedPost as unknown as Post;
}
```

This will create a post with the submitted information and the authenticated user as the author.

11. Implement the `comment` mutation, as follows:

```
comment: async (_, args, { orm, authUser }: Context) => {
  const comment = orm.commentRepository.create({
    comment: args.comment,
    post: await orm.postRepository.findOne(
      args.postId),
    author: await orm.userRepository.findOne(
      authUser?.id)
```

```
    } as CommentEntity);

    const savedComment = await
      orm.commentRepository.save(comment);
    await orm.postRepository.update(args.postId, {
      commentsCount: savedComment.post.commentsCount +
        1,
      latestComment: savedComment
    });
    savedComment.post = await
      orm.postRepository.findOne(args.postId) as
      PostEntity;
    return savedComment as unknown as Comment;
  }
```

12. Next, implement the `like` mutation, as follows:

```
like: async (_, args, { orm, authUser }: Context) => {
  const like = orm.likeRepository.create({
    user: await orm.userRepository.findOne(
      authUser?.id),
    post: await orm.postRepository.findOne(
      args.postId)
  } as LikeEntity);
  const savedLike = await
    orm.likeRepository.save(like);
  await orm.postRepository.update(args.postId, {
    likesCount: savedLike.post.likesCount + 1
  });

  savedLike.post = await orm.postRepository.findOne(
    args.postId) as PostEntity;
  return savedLike as unknown as Like;
}
```

13. Next, implement the `removeLike` method, as follows:

```
removeLike: async (_, args, { orm, authUser }: Context)
=> {
  const like = await
    orm.likeRepository.createQueryBuilder("like")
    .innerJoinAndSelect("like.user", "user")
    .innerJoinAndSelect("like.post", "post")
    .where("post.id = :postId", { postId: args.postId
    })
    .andWhere("user.id = :userId", { userId:
      authUser?.id })
    .getOne();
  if (like && like.id) {
    const result: DeleteResult = await
      orm.likeRepository.delete(like.id);
    if (result.affected && result.affected <= 0) {
      throw new ApolloError("Like not deleted",
        "LIKE_NOT_DELETED");
    }
  }
  if (like && like.post && like.post.likesCount >= 1)
  {
    await orm.postRepository.update(like.post.id, {
      likesCount: like.post.likesCount - 1
    });
  }
  return like as unknown as Like;
}
```

14. We also need to update the getPostsByUserId and getFeed queries to set the likedByAuthUser field. Let's start with the getPostsByUserId resolver:

```
posts.forEach((post: PostEntity) => {
  if (post.likes?.find((like: LikeEntity) => { return
    like.user.id == ctx.authUser?.id })) {
    post.likedByAuthUser = true;
  }
});
return posts as unknown as Post[];
```

Just above the return statement, we iterate over the posts array and check whether the current user has liked the post. If so, we simply set the likedByAuthUser property of the post to true.

For the getFeed resolver, add the following:

```
feed.forEach((post: PostEntity) => {
  if (post.likes?.find((like: LikeEntity) => { return
    like.user.id == ctx.authUser?.id })) {
    post.likedByAuthUser = true;
  }
});
return feed as unknown as Post[];
```

Check out the GitHub commit at https://git.io/JgCDh.

15. Now, let's experiment with these mutations using Apollo Studio. Let's start by creating some posts using the following mutation:

```
mutation CreatePosts {
  post1: post(text: "This is post 1"){
    id author { id fullName }
  }
  post2: post(text: "This is post 2"){
    id author { id fullName }
  }
  post3: post(text: "This is post 3"){
    id author { id fullName }
  }
}
```

This is the response I received:

```
Response ∨  ≕  ⊞                              ● STATUS 200  122ms  215B

{
  "data": {
    "post1": {
      "id": "2",
      "author": {
        "id": "1",
        "fullName": "Ahmed Bouchefra"
      }
    },
    "post2": {
      "id": "3",
      "author": {
        "id": "1",
        "fullName": "Ahmed Bouchefra"
      }
    },
    "post3": {
      "id": "4",
      "author": {
        "id": "1",
        "fullName": "Ahmed Bouchefra"
      }
    }
  }
}
```

Figure 4.4 – The response of the post mutation

You can see that the author is the same as the authenticated user that we used when signing in using the `signIn` mutation. It was retrieved successfully from the token we passed via the authorization header. This gives us a sign that our implementation is working as expected.

Next, let's create some comments and likes on these posts:

```
mutation CreateComments {
  c1: comment(comment: "Comment on post with id #2",
    postId: 2 ) {
    id comment author { id fullName } post { id }
  }
  c2: comment(comment: "Comment on post with id #3",
    postId: 3 ) {
    id comment author { id fullName } post { id }
  }
  c3: comment(comment: "Comment on post with id #4",
    postId: 4 ) {
```

```
        id comment author { id fullName } post { id }
    }
}
```

In the response, the `post` ID that you receive in the selection set should be the same as the one you pass to the mutation's `postId` parameter. This means that we are correctly retrieving relations in our resolver. Also, the author in the selection set should be the same as the authenticated user.

Let's create likes in the same way:

```
mutation CreateLikes {
    like1: like(postId: 1) {
        id user { fullName } post { id }
    }
    like2: like(postId: 2) {
        id user { fullName } post { id }
    }
    like3: like(postId: 3) {
        id user { fullName } post { id }
    }
}
```

We have protected our GraphQL endpoints and only allowed authenticated users to interact with most queries and mutations of the API. We also implemented the `post`, `comment`, `like`, and `removeLike` mutations. Next, we'll see how to implement image uploading with Apollo Server to allow users to upload their photos and cover images for their profiles.

Implementing image uploads

After implementing authentication in our GraphQL API and protecting the queries and mutations, let's now implement image uploading that will be used to upload the user's profile photo and cover image.

We'll see how to implement file uploading with Apollo Server and stream the images to storage services, such as Amazon S3 or DigitalOcean Spaces Object Storage. We'll be using this last service, but feel free to use Amazon S3 if you prefer that or any S3-compatible object storage service.

Before that, make sure to sign up for your chosen S3 compatible service, create an S3 bucket, and take note of the following information:

- S3_ACCESS_KEY_ID
- S3_SECRET_ACCESS_KEY
- S3_ENDPOINT
- S3_BUCKET

Using the server URL, bucket name (or Space name), access key, and secret, you can connect any S3-compatible client or library to your DigitalOcean Space. A Space is equivalent to a bucket.

The DigitalOcean Spaces API is interoperable with the Amazon S3 REST API, so we'll be using AWS SDKs for interacting with Spaces:

1. Open the `server/src/graphql/schema.graphql` file and add the following line at the top:

    ```
    scalar Upload
    ```

2. Then, add the following type:

    ```
    type File {
        url: String
        filename: String
        mimetype: String
        encoding: String
    }
    ```

3. Next, add the following mutation:

    ```
    type Mutation {
        [...]
        uploadFile(file: Upload!): File!
    }
    ```

4. Next, you need to head back to your Terminal and run the following commands from the server's folder:

    ```
    npm i graphql-upload aws-sdk
    npm i --save-dev @types/graphql-upload
    ```

This will install `graphql-upload` and its types, plus the `aws-sdk` packages.

5. Next, open the `server/src/index.ts` file and start by adding this `import`:

```
import { graphqlUploadExpress } from 'graphql-upload';
```

Then, use the middleware, as follows:

```
async function startApolloServer() {
    const PORT = 8080;
    const app: Application = express();
    app.use(cors());
    app.use(graphqlUploadExpress());
    // [...]
```

Make sure to use the `graphqlUploadExpress` middleware before calling the `server.applyMiddleware()` method.

We have now enabled file uploading with Apollo Server and Express; next, we'll need to implement the `uploadFile` resolver.

Before that, open the `.env` file and add the following variables with your own S3 service credentials:

```
S3_ACCESS_KEY_ID=<YOUR_ACCESS_KEY>
S3_SECRET_ACCESS_KEY=< YOUR_SECRET_KEY >
S3_ENDPOINT=<YOUR_S3_ENDPOINT>
S3_BUCKET=<YOUR_S3_BUCKET>
```

Also, make sure to regenerate your resolvers' types (using `npm run codegen` from the root of your Lerna project).

6. The implementation of the `Upload` scalar is provided by the `GraphQLUpload` export from the `graphql-upload` package, so you need to add it to your resolvers' map to enable it. Open the `server/src/graphql/resolvers.ts` file and add the following import:

```
import { GraphQLUpload } from 'graphql-upload';
```

Then, add it to the `resolvers` object:

```
const resolvers: Resolvers = {
    Upload: GraphQLUpload,
    Query: {...},
    Mutation: {...}
};
```

Next, add the following code at the top:

```
import AWS from 'aws-sdk';
const spacesEndpoint = new AWS.Endpoint(process.env.S3_
ENDPOINT as string);
const s3 = new AWS.S3({
  endpoint: spacesEndpoint,
  accessKeyId: process.env.S3_ACCESS_KEY_ID,
  secretAccessKey: process.env.S3_SECRET_ACCESS_KEY
});
```

Finally, add the `uploadFile` mutation, as follows:

```
uploadFile: async (_, { file }) => {
  const { createReadStream, filename, mimetype,
    encoding } = await file;
  const fileStream = createReadStream();
  const uploadParams = { Bucket: process.env.S3_BUCKET
    as string, Key: filename, Body: fileStream, ACL:
    "public-read" };
  const result = await
    s3.upload(uploadParams).promise();
  console.log(result);
  const url = result.Location;
  return { url, filename, mimetype, encoding };
}
```

We destructure the `file` promise passed as an argument to the resolver. This way, we can get the read stream and other information, such as the filename, MIME type, and encoding needed to save the file to the S3 service using the `upload()` method of the AWS SDK. Next, we get the URL of the saved file and we return with the other information.

Assignments

We'll use the `uploadFile` mutation to upload images for posts. For uploading the user's photo and cover image, we also need to link the uploaded images to the appropriate user. To achieve this, we need to add two `setUserPhoto` and `setUserCover` mutations.

Follow the same steps for adding the `setUserPhoto` and `setUserCover` mutations to the schema. Both mutations should have a `file` argument, of the `Upload` type, and should return a `User` object.

Next, regenerate the resolvers' types and then implement the required resolvers.

Finally, use the `@auth` directive to protect the mutations from unauthenticated users.

You can find the solution for this assignment at `https://git.io/JgQiC`.

As a second assignment, add the `setUserBio` mutation (which takes a `bio` argument of the `String` type and returns the updated `User` object), regenerate the resolvers' types, and implement the mutation's resolver.

You can find the solution for this assignment in the commit at `https://git.io/JgQzA`.

Summary

In this chapter, we saw how to add authentication and image uploading to our GraphQL API with JWT, and we protected the endpoints from users that are not authenticated.

That concludes this chapter! In the next chapter, we'll continue our journey of building our backend application by implementing realtime subscriptions.

5
Adding Realtime Support with Apollo Server

In the previous chapter, we saw how to add authentication and image uploading to our **GraphQL** API. We've also seen how to protect endpoints from access by non-authenticated users.

Traditionally, in web applications, clients send HTTP requests to servers and receive the response. After that, the connection gets closed. As a result, if the data on the server is changed, the client doesn't have the mechanism to know that.

Nowadays, contemporary apps operate in real time, which means that the backend server can alert clients anytime data changes, even if they do not send another request, because the connection remains open to allow the server to transmit updates whenever they become available.

In this chapter, we'll add realtime support to our server application, which will allow us to communicate fresh data from the server to the client as soon as it becomes available. To do this, we'll leverage Apollo Server's GraphQL subscriptions.

We will address the following topics:

- Understanding GraphQL subscriptions
- Implementing GraphQL subscriptions with Apollo Server
- Implementing **JSON Web Token** (**JWT**) authentication with subscriptions

Technical requirements

To accomplish the steps in this chapter, you must have completed the previous chapter's steps. Refer to *Chapter 4, Implementing Authentication and Image Uploads with Apollo Server*, of this book for instructions.

You also need to be familiar with the following technologies:

- TypeScript
- Node.js
- GraphQL basics

You can find the code source for this chapter in the `chapter5` folder of our GitHub repository or via this link: at `https://github.com/PacktPublishing/Full-Stack-App-Development-with-Angular-and-GraphQL/tree/main/Chapter05`.

Understanding GraphQL subscriptions

When interacting with a GraphQL server, you'll often get and write data using queries and mutations. In this situation, the client must send a query or mutation to the server in order to obtain a response.

Subscriptions are another form of data communication in GraphQL, in which the server provides realtime data to listening clients.

In order to do that, the client opens a two-way communication with the server, which can be done with protocols such as WebSocket instead of HTTP. WebSocket can be used to make two-way communication between the user's browser and a server.

Subscriptions are similar to queries. However, they let subscribing clients receive fresh or updated data whenever a certain server event happens. You may configure Apollo Server to use a subscription-specific endpoint that is distinct from the default endpoint for queries and mutations.

We will be able to add realtime capabilities to our social network application thanks to GraphQL subscriptions. We may create them in the same way that we define queries, with a group of fields indicating the form of data required by the client, but under a `Subscription` type in our schema.

Now that we've defined GraphQL subscriptions, we'll see how to use them with Apollo Server to provide realtime capabilities to our social network application.

Implementing GraphQL subscriptions

After explaining GraphQL subscriptions, let's look at how we can use Apollo Server to implement them in our application. Because subscriptions utilize WebSocket rather than HTTP, we must use a second GraphQL endpoint created for subscriptions that uses the WebSocket protocol rather than HTTP:

1. Begin by installing the packages listed as follows:

    ```
    npm install subscriptions-transport-ws graphql-
    subscriptions
    ```

2. Add the following imports to the `server/src/index.ts` file:

    ```
    import { createServer } from 'http';
    import { execute, subscribe } from 'graphql';
    import { SubscriptionServer } from 'subscriptions-
       transport-ws';
    ```

3. Following that, we must launch two servers, HTTP and WebSocket. As a result, we must create `http.Server` by providing our Express app to the `createServer()` function in the body of the `startApolloServer()` function, as seen here:

    ```
    async function startApolloServer() {
      const PORT = 8080;
      const app: Application = express();
      // [...]
      const httpServer = createServer(app);
    ```

4. Next, create a `SubscriptionServer` instance beneath the `ApolloServer` instance:

```
const subscriptionServer = SubscriptionServer.create(
    { schema, execute, subscribe },
    { server: httpServer, path: server.graphqlPath }
);
```

5. Then, in the `ApolloServer` constructor, add a plugin to close the `SubscriptionServer` instance.

    First, add the `plugins` array to the constructor, as follows:

```
const server: ApolloServer = new ApolloServer({ schema,
context: ({req}) => {
    // [...]
    return ctx;
}, plugins: [] });
```

Then, inside the `plugins` array, add the following code:

```
{
    async serverWillStart() {
        return {
            async drainServer() {
                subscriptionServer.close();
            }
        };
    }
}
```

6. Update the `listen()` call from `app.listen()` to `httpServer.listen()`:

```
httpServer.listen(PORT, () => {
    console.log('Server is running at
        http://localhost:${PORT}');
});
```

This will allow us to launch both servers. Check out this commit for the updates we've made in this step: `https://git.io/JghrI`.

7. Next, we need to update the schema, open the `server/src/graphql/schema.graphql` file, and add the following type:

```
type Subscription {
  onPostCommented: Comment
  onPostLiked: Like
}
```

We added a `Subscription` type that defines the top-level fields that clients can subscribe to.

The `onPostCommented` field will update its value whenever a new comment is created on the backend and send the comment to subscribing clients.

Before continuing, you need to regenerate the resolvers' types using the `npm run codegen` command from the root of your monorepo project.

Next, we need to implement the resolvers for these two fields. Resolvers for these fields are different from the resolvers for the query and mutation fields. They are not functions but objects that should define a `subscribe` function. This function must return an object of the `AsyncIterator` type, a standard interface for iterating over asynchronous results:

1. Open the `server/src/graphql/resolvers.ts` file and start by adding the following code:

```
import { PubSub } from 'graphql-subscriptions';
const pubsub = new PubSub();
```

We imported the `PubSub` interface and created an instance that will allow you to publish events to a particular label and listen for events associated with a particular label.

Apollo Server adopts a publish-subscribe mechanism to track events for updating active subscriptions.

2. Next, add a `Subscription` object inside the `resolvers` object, as follows:

```
Subscription: {
  onPostCommented: {
    subscribe: () =>
      pubsub.asyncIterator(['ON_POST_COMMENTED'])
  },
  onPostLiked: {
    subscribe: () =>
```

```
      pubsub.asyncIterator(['ON_POST_LIKED'])
    }
  }
```

For each field, we added a `subscribe()` function and mapped the corresponding event using the `pubsub.asyncIterator()` method.

3. Next, you need to publish some events with the `publish()` method of the PubSub object.

    Update the `comment` mutation, as follows:

    ```
    comment: async (_, args, { orm, authUser }:
      Context) => {
      // [...]
      pubsub.publish('ON_POST_COMMENTED',
      { onPostCommented: savedComment });
      return savedComment as unknown as Comment;
    }
    ```

    Then, update the `like` mutation, as follows:

    ```
    like: async (_, args, { orm, authUser }: Context) => {
      // [...]
      pubsub.publish('ON_POST_LIKED',
      { onPostLiked: savedLike });
      return savedLike as unknown as Like;
    }
    ```

    Apollo Server uses the payloads of the ON_POST_COMMENTED and ON_POST_LIKED events to push updated values for the onPostCommented and onPostLiked fields respectively.

4. We can write the GraphQL query for listening for subscriptions in the same way we write other types of queries. Head over to `http://localhost:8080:graphql` and write the following subscription in the left panel of Apollo Studio:

    ```
    subscription {
      onPostCommented {
        comment
        post {
          id
        }
    ```

```
    }
  }
```

After running this subscription, we will listen for comments on posts over WebSocket. In Apollo Studio, this will not return data immediately like queries and mutations. Instead, we'll get data every time a comment is added.

If you want to see some data pushed to an `onPostCommented` subscription, you simply need to create a comment.

You can start listening for new likes using the following subscription:

```
subscription {
  onPostLiked {
    id
    post {
      id
      likesCount
    }
  }
}
```

This is an example of a response I received after liking the post with the 1 ID:

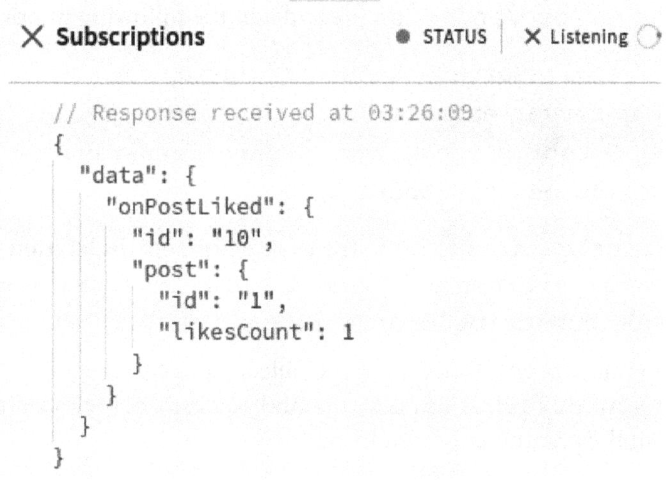

```
X Subscriptions                    ● STATUS   X Listening ◯

// Response received at 03:26:09
{
  "data": {
    "onPostLiked": {
      "id": "10",
      "post": {
        "id": "1",
        "likesCount": 1
      }
    }
  }
}
```

Figure 5.1 – The response of a subscription

Now that we have implemented subscriptions to add realtime features to our social application, let's see how to secure these subscriptions using JWT authentication.

Implementing JWT authentication with subscriptions

This section explains how to implement authentication over WebSocket. Refer to the documentation at `https://www.apollographql.com/docs/apollo-server/data/subscriptions/#operation-context` for more information.

Until now, our subscriptions have not been secure, so we need to prevent subscriptions from users that are not authenticated. This means that we only allow WebSocket connections if the user is authenticated.

We can secure our subscriptions in the same way we did with queries and mutations. We simply need to pass a context parameter to our subscriptions each time a user connects over WebSocket!

When we create an instance of `SubscriptionServer`, we can use an `onConnect` function that gets executed before every WebSocket connection. This function accepts an object of the `ConnectionParams` type as one of its arguments. If it returns an object, it gets passed to the resolvers as context.

Using `ConnectionParams`, we can get the JWT sent from the client. Inside `onConnect`, we will extract the authenticated user from the JWT and add it to the context:

1. Open the `server/src/index.ts` file and add the following imports:

    ```
    import { ApolloServer, AuthenticationError } from
      'apollo-server-express';
    import { SubscriptionServer, ConnectionParams } from
      'subscriptions-transport-ws';
    ```

    We add `AuthenticationError` to the existing import declaration from `apollo-server-express` and `ConnectionParams` from the existing import declaration from `subscriptions-transport-ws`.

2. Next, add the `onConnect` function to the object passed to the `SubscriptionServer.create()` method and extract the token that can be retrieved from the parameter passed to `onConnect`:

    ```
    const subscriptionServer = SubscriptionServer.create(
      {
        schema, execute, subscribe, onConnect:
          (connectionParams: ConnectionParams) => {
    ```

```
        const token = connectionParams.get('authToken')
          || '';
      if (token != '') {
        const authUser = getAuthUser(token.split('
          ')[1]);
        return {
          authUser: authUser
        }
      }
      throw new AuthenticationError('User is not
        authenticated');
    }
  },
  { server: httpServer, path: server.graphqlPath }
);
```

We've now learned how to secure our WebSocket connection to prevent users that are not authenticated from subscribing to realtime events.

Summary

Throughout this chapter, we learned how to add realtime support to our server application to push new comments and likes on users' posts from the server to the client, right at the moment when they are added. To achieve this, we used GraphQL subscriptions with Apollo Server.

Next, we implemented JWT authentication to prevent users that are not authenticated from subscribing to these realtime subscriptions.

Now that we have a working GraphQL API, with realtime subscriptions and JWT authentication, in the next chapter we'll start implementing our Angular frontend for building the UI of our social network application.

Part 2: Building the Angular Frontend with Realtime Support

In this part, we'll implement the frontend of our application using Angular and Apollo. We'll add routing and navigation between various modules and components.

We'll look at how to integrate the frontend with the backend using Apollo Client, which is designed for sending GraphQL queries and mutations to the server to fetch and write data, and then we'll implement the authentication system.

Following that, we will implement the profile component's functionality by adding the ability to upload the user's photo and cover image, as well as a biography.

We'll add the necessary code to fetch the user that corresponds to a profile URL and render their information on the page and then we'll learn about sending queries to receive paginated posts and comment-related data and mutations to add comments and likes to posts.

This section comprises the following chapters:

- *Chapter 6, Angular Application Architecture and Routing*
- *Chapter 7, Adding User Search Functionality*
- *Chapter 8, Guarding Routes and Testing Authentication*
- *Chapter 9, Uploading Images and Adding Posts*
- *Chapter 10, Fetching Posts and Adding Comments and Likes*

6
Angular Application Architecture and Routing

In earlier chapters, we developed our backend using GraphQL and Apollo Server, and we also used Apollo Studio to communicate with our GraphQL API. We now can begin developing our frontend application to consume the API and present a user interface to users to communicate with the backend.

In the following chapters, we'll focus on connecting our frontend application with the backend API through Apollo Client and sending queries for obtaining and mutating data, but first we'll need to build an Angular frontend project with a basic layout.

In this chapter, we'll begin by installing the **Angular CLI** and creating a new project using a recent version of Angular. Following that, we'll utilize the Angular CLI to create the **modules**, **services**, and **components** that make up our application's UI, as well as introducing you to **dependency injection**.

Following the creation of the various artifacts required for our project, we will introduce you to **Angular Router** and demonstrate how to implement routing and navigation to develop a single-page application. Following that, we'll walk you through debugging using the current **Angular Ivy** runtime, which will allow you to access your application components from the console, execute methods, and even trigger change detection. Finally, we'll install and configure **Angular Material** in our project to provide components for developing the user interface of our application.

The following topics will be covered:

- Installing the Angular CLI
- Initializing the Angular project
- Understanding the application architecture
- Understanding services and dependency injection
- Creating modules, services, and components
- Understanding Angular routing
- Understanding and adding routes
- Adding navigation
- Debugging Angular applications
- Adding Angular Material

Technical requirements

You must have Node.js and npm installed on your local development machine in order to follow the instructions in this chapter. Refer to *Chapter 1*, *App Architecture and Development Environment*, of this book for instructions on how to install them if you haven't already.

You must also finish the previous chapters in order to develop a GraphQL API that will be used by the frontend that we will be implementing beginning with this chapter.

You should also be familiar with the following technologies:

- JavaScript/TypeScript
- HTML

You can find the source code of this chapter at `https://github.com/ PacktPublishing/Full-Stack-App-Development-with-Angular-and- GraphQL/tree/main/Chapter06`.

TypeScript expertise is not required, however, knowledge of concepts such as classes and decorators is helpful.

An example of a TypeScript class is as follows:

```
class Product {
  name: string;
  constructor(name: string) {
    this.name = name;
  }
  getName() {
    return this.name;
  }
}

let product1 = new Product("product1");
```

If you've worked with a class-based oriented programming language before, you should be familiar with this syntax, but even recent versions of JavaScript, such as ECMAScript 6 or JavaScript 2015, enable developers to define classes using this syntax.

We simply define classes by using the `class` keyword, followed by the name and the body of the class.

The members of the class are added to the body, which is surrounded by braces. In this case, we have a property named `name` of the `string` type, a constructor, and a method named `getName()`.

Finally, we create an instance of the `Product` class using the new keyword, which creates a new object with the `Product` type and invokes the constructor of the class to initialize it.

Classes may be thought of as a mechanism to create user-defined types that allow you to easily represent real-world objects.

In TypeScript, a decorator is a special construct that allows you to alter classes, methods, accessors, properties, and arguments dynamically.

> **Important Note**
>
> More information regarding classes may be found in the TypeScript handbook at `https://www.typescriptlang.org/docs/handbook/classes.html`. You can refer to `https://www.typescriptlang.org/docs/handbook/decorators.html` for additional information on decorators.

This is an example of a class that has been decorated with the `@Component` decorator, which is available from the Angular core module and is used to create components from TypeScript classes:

```
@Component({
  selector:    'product-list',
  templateUrl: './product-list.component.html'
})
export class ProductListComponent implements OnInit {
  products: Product[];
  myProduct;
  constructor(private service: ProductService) { }
  ngOnInit() {
    this.products = this.service.getProducts();
  }
  selectProduct(product: Product) {
    this.myProduct = product;
  }
}
```

Later in this chapter, we'll look at what a component is and what the different attributes supplied to the `@Component` decorator do.

Angular heavily relies on custom decorators accessible from the core module, such as `@Component`, `@Injectable`, and `@NgModule`, to generate a variety of artifacts that may be used to structure applications.

Installing the Angular CLI

In this section, we'll learn how to install the Angular CLI, which will allow us to quickly get started with our Angular project without having to deal with the complexity of setting **webpack** or another build system.

The official utility for creating Angular projects is the Angular CLI. It simplifies the process of creating fully functional Angular projects.

The Angular CLI allows you to do more than just create project files and install dependencies. It also helps you throughout the development process, from serving your application locally through testing and generating final production bundles and deploying them to your chosen hosting provider.

Now, let's begin with the chapter's first practical step. By opening a new command-line interface and entering the following command, you can install the Angular CLI from npm:

```
npm install -g @angular/cli
```

At the time this chapter was written, `angular/cli` version 12 is installed on our development machine.

> **Important Note**
> In order to globally install packages on your machine, you may need to include `sudo` before your command in your Ubuntu system. Check out `https://www.shabang.dev/multiple-versions-node-nvm`.

Angular CLI commands

The Angular CLI includes a plethora of commands to aid you in the development of your project. Simply running the `ng` executable in a command-line interface will display the available commands along with their explanations. The following are some of the most frequently used commands:

- The `ng new` command is used to create an application.
- The `ng generate` command is used to create Angular constructs like modules, components, and services.
- The `ng serve` command is used for starting a live-reload server during the development of your project.
- The `ng build` command is used to create your project's final bundles.

We installed the Angular CLI, and went through some of the CLI's most useful commands. In the next section, we'll look at how to utilize the CLI to create and serve a project.

Initializing the Angular project

This section describes how to create our Angular project as well as how to build, monitor, and serve the application.

Return to your command-line interface and go to your project's `packages/` subdirectory:

```
cd ~/ngsocial/packages/
```

Then, to create a new Angular project without a Git repository, use the following command:

```
ng new client --skip-git
```

The CLI will ask you a couple of questions:

- `Would you like to add Angular routing?` Type `Y` for yes.
- `Which stylesheet format would you like to use?` Choose `CSS`.

> **Important Note**
>
> To see what version of the Angular CLI we're using, we can run the `ng version` command when we're within our project's folder. Prompt questions can be skipped by using the `--routing` and `--style=css` options with the `ng new` command telling the Angular CLI that you want to create a project using pre-configured routing and CSS stylesheets.

Following confirmation, the CLI will create the basic folder structure and source files automatically, install the required dependencies from npm, and *since you have said yes for Angular routing, it will also set up routing in your project.*

The above command will create a **workspace** with an initial application named after your project and located in the project's root folder. A workspace can include many apps and libraries.

Your application source code exists within the `src/` subdirectory, which is where the TypeScript source files are created.

You'll see several configuration files outside the `src/` subdirectory, including the following:

- The `package.json` file for npm.
- The `tsconfig.json` file for configuring TypeScript.

- `karma.conf.js` for the **karma** test runner.

- `angular.json` for Angular CLI configuration.

Following that, navigate to the folder containing your project and launch the local development server using the following commands:

```
cd client
ng serve
```

The local live-reload development server will be accessible through the `http://localhost:4200/` address. When you make changes to the code, your application is rebuilt and delivered to the browser automatically, without you having to refresh it manually.

Open your web browser and navigate to `http://localhost:4200/` to see your Angular application running with some initial contents, which we'll erase before creating the demo's actual UI. The following is a screen capture of the page:

Figure 6.1 – Angular project default page

Now that we've created an Angular project and started a live-reload development server, we're ready to go on to the following stages, which include an introduction to Angular's fundamental ideas, such as modules, components, and services, followed by the creation of the user interface.

Understanding the application architecture

Throughout this book, we'll be developing a full stack web application with the functionality of a small social network. We already discussed the architecture of the backend application, which utilizes Node.js, Apollo Server, and GraphQL. Now, we're going to create the frontend of our application using Angular and Apollo Client to consume the server's GraphQL API.

Angular is a Google-developed platform for creating client-side web apps for mobile and desktop using TypeScript. We'll use it to consume the server's GraphQL API and build the application's user interface. Additionally, we'll utilize Angular Material to style the UI.

TypeScript is a Microsoft-created superset of JavaScript that adds static typing, improved error checking, type safety, and object-oriented programming principles to the language before it is compiled to JavaScript and processed in web browsers. Additionally, it integrates more tightly with code editors and integrated development environments (IDEs), which enhances the developer experience, particularly when developing big JavaScript applications that must scale.

In comparison to other popular JavaScript libraries, such as React, Angular is a framework that includes built-in libraries for common development tasks, such as the following:

- **HttpClient** to communicate with servers using HTTP
- **Angular Universal** to render your Angular application on the server
- **Angular Material** to provide material components for styling the user interface
- **Angular forms** to create, validate, and manipulate forms

Angular supports both modular and component-based architectures, which enable you to build your application using modules and components.

When developing an Angular application, you may break it down into modules and components that can be reused across the application and, with care, across numerous applications.

While the Angular APIs are contained within built-in modules, you can also develop custom modules to structure and organize your application. A module can include components, services, and any other Angular artifacts. Let's take a closer look at modules and components.

What's a module?

A module is nothing more than a container for Angular code that implements a certain domain function. They allow you to break up your application into smaller, more manageable pieces.

In order to avoid having to write all of the code from scratch for every new project, the code units may be combined into libraries that can be used by other developers.

An Angular module can include components, directives, pipes, and services, and can be lazy-loaded by the router to improve application speed.

You may either design your own custom modules to implement a requirement or feature in your app, or you can utilize the built-in modules that expose the Angular APIs, such as the ones listed here:

- `FormsModule` for creating and validating forms.
- `HttpClientModule` for making HTTP requests.
- `RouterModule` for supplying your Angular application with a router as well as a variety of helpful APIs and components.
- `CommonModule` for supplying all of the basic Angular directives and pipes, such as `NgIf` and `NgForOf`.

NgModules may be used to define Angular modules. `NgModule` is a TypeScript class annotated with the `@NgModule` decorator that accepts certain metadata and performs the following functions:

- Declares other module artifacts, such as components, directives, and pipes
- Exports some components, directives, or pipes so that other modules can import and use them
- Imports other modules with their components, directives, and pipes
- Provides services that the other components can use

By convention, an Angular application comprises at least one module, which is referred to as the root module. This is the module that is used to get the application up and running.

In our Angular CLI-generated project, we already have two modules included in the `src/app/app.module.ts` and `src/app/app-routing.module.ts` files, which export the classes used to define the root application module and the routing module.

A module may be made up of one or more components. So, what exactly is a component?

What's a component?

An **Angular component** is in charge of displaying an area of the application's user interface. A component, for example, may be used to make your application's home page, and if you want to gain further code reusability, you can further break the home page into smaller components, such as the navigation bar, main, and footer areas.

This allows you to incorporate the navigation and footer components on other pages of your application without having to rewrite the same code for the elements that are shared by all, or just some, parts of your application.

In terms of code, a component is just a TypeScript class that has been decorated with the @Component decorator exported from the Angular core module. It also includes an HTML template, which may be inline or saved as a separate HTML file and uses specific Angular template syntax to show data and bind properties and events from the component's class.

Finally, a component may have one or more associated style sheet files that include styles for the view in the format you've selected for your project, such as CSS, SCSS, or Sass.

Angular 9 arrives with a new rendering engine called Ivy, which is considered a next-generation compilation and rendering system, consisting of a compiler and a set of runtime instructions.

Ivy yields smaller production bundles, faster build times and runtime performance, and easier debugging. It makes use of **incremental DOM** where Angular components and templates are compiled into a group of JavaScript instructions that are used to create and update the **Document Object Model (DOM)**.

Let's get back to our application for a moment. In the src/app folder, you'll find a number of files that include the TypeScript code, HTML template, and CSS style sheet for the app component:

- The app.component.css file that provides the CSS styles for the component
- The app.component.html file that provides the HTML code for rendering the component
- The app.component.ts file that provides the TypeScript code for defining the component data and functionality

When the app is launched, the root component is first loaded. This is accomplished by the inclusion of the component in the application module's bootstrap array.

Simply open the `src/app/app.module.ts` file containing the application module to see this:

```
// [...]
import { AppComponent } from './app.component';

@NgModule({
  // [...]
  bootstrap: [AppComponent]
})
export class AppModule { }
```

The root component is also included in the `src/index.html` file, which is the file that gets downloaded when users visit your Angular application from a web browser:

```
<!doctype html>
<html lang="en">
<head>
  <!-- [...] -->
</head>
<body>
  <app-root></app-root>
</body>
</html>
```

Through its selector, you can see that the root component is being utilized exactly like a standard HTML tag. This instructs Angular to load the component and all of its descendant components into the DOM.

Open the `app.component.ts` file in the `src/app` directory. The root component should be defined using the following code:

```
import { Component } from '@angular/core';

@Component({
  selector: 'app-root',
  templateUrl: './app.component.html',
  styleUrls: ['./app.component.css']
})
export class AppComponent {
```

```
    title = 'client';
}
```

This is just an ordinary TypeScript class that has the `@Component` decorator, imported from the core module, applied to it. The decorator contains metadata about the component, such as the following:

- `selector`: This is used in the same way as conventional HTML tags to call the component from an HTML page. You may use Angular to build custom HTML tags that are only useable within your application.

- `templateUrl`: This is used to tell Angular where to look for the HTML template that contains the presentation markup.

- `styleUrls`: This is used to designate where to find component-specific styles that determine the appearance and feel of the component.

A component's life cycle begins when Angular constructs an instance of the component's class and ends when the component is destroyed and removed from the DOM.

Angular supports lifecycle hook methods via a slew of interfaces, each of which contains a method, to allow you to hook onto important events in a component's lifecycle.

The `OnInit` interface, for example, provides the `ngOnInit` method, which is run when the component is initialized. Please refer to `https://angular.io/guide/glossary#lifecycle-hook` for further information about lifecycle hooks.

Now that we've defined modules and components, let's look at services and dependency injection.

Understanding services and dependency injection

A service in Angular is a singleton that may be provided to other components or services using `dependency injection`, which is a way to supply dependencies to an object.

Don't be misled by this terminology! This simply means that an Angular entity called the **injector** is in charge of creating an instance of the service and delivering it to the components that request it.

Dependency injection is tightly linked with the Angular framework and is used to inject services and other objects into components, allowing access to the service functionality without having to manually create an instance.

Angular services enable you to create apps that adhere to the following guidelines:

- Isolation of your application's business logic from its display logic in components
- Interchange of data between components in your application
- Reusability of code

This means that you should leverage services to organize and reuse code within your application in order to prevent duplication and reinventing the wheel.

Generally, you should utilize components to create your application's user interface by supplying properties and methods to the HTML template (which is responsible for creating the view) via data binding and events. You can refer to `https://angular.io/guide/binding-syntax` for more information about Angular binding syntax.

On the other hand, services are necessary for common tasks that several components require to function, such as data retrieval from the server.

In Angular, you may define a service by using the `@Injectable` decorator, which allows you to supply the information that the framework needs to inject the service as a dependency into a component or other service.

How do we then instruct Angular to inject a service?

By simply passing the dependency as a parameter to the constructor of the requesting component. This is sufficient for Angular to provide the component with a service instance. However, before Angular can accomplish that, we must supply the service in one of the following ways:

- By adding the service to a component's `providers` array, it becomes exclusive to that component.
- By including the service in a module's `providers` array, it becomes available solely within the module.
- By utilizing the `providedIn` property in conjunction with the `@Injectable` decorator. For further details, see `https://next.angular.io/api/core/Injectable`.

Now that we've defined modules, components, and services, we'll explore how to create them using the Angular CLI.

Creating modules, services, and components

You may automatically create components, services, and modules in Angular by calling the relevant Angular CLI command, which generates the bare minimum code required for a basic component, service, or module.

Let's use the Angular CLI to generate our modules, services, and components.

Create a module for each feature of your application. This is a basic rule to follow. For instance, the home feed can be contained within a separate module. Splitting your application into modules enables you to lazily load components, and maintain it efficiently.

Creating modules

In our application, we'll utilize the following modules:

- A `users` module that incorporates components for login, signup, and profile.

- A `feed` module that includes the component(s) necessary to show the posts retrieved using GraphQL.

- A `shared` module may include elements that are shared with other modules in your application.

- A root module containing our application's global services and common layout components, such as the authentication service and header and footer components.

Shared modules save you time by avoiding the import of common modules, components, and other artifacts that are required by several modules in your application.

For instance, by importing and exporting `ReactiveFormsModule` from our shared module, we can avoid importing it in both the `users` and `feed` modules by simply importing the shared module.

Additionally, we'll import and re-export Angular Material component modules into the shared module to make them available to the other modules in our project.

Let's begin by creating the modules automatically using the Angular CLI. Open a new Terminal window, browse to the folder containing your Angular project, and run the following commands:

```
ng generate module shared
ng generate module users --routing
ng generate module feed --routing
```

These commands will create a number of files under the `src/app/` folder, each with its own subdirectory called after the module we specified.

The shared module, for example, will be created under the `src/app/shared/shared.module.ts` file. Open the file and make the following changes:

```
import { NgModule } from '@angular/core';
import { CommonModule } from '@angular/common';
import { ReactiveFormsModule } from '@angular/forms';

@NgModule({
  declarations: [],
  imports: [
    CommonModule,
    ReactiveFormsModule
  ],
  exports:[
    ReactiveFormsModule
  ]
})
export class SharedModule { }
```

We simply imported and added `ReactiveFormsModule` to the `imports` array, and then re-exported it via the `exports` array. This makes it available to all modules that import the `shared` module, eliminating the need to import it separately.

The `--routing` option instructs the CLI to create a routing module for each generated module, allowing us to implement routing for each module's components. Additionally, it will enable us to use lazy loading, which enhances the speed of our application by allowing Angular to load modules only when they are required, rather than loading them all at startup.

> **Important Note**
>
> Additionally, we may add `CommonModule` to the `exports` array to re-export it from the `shared` module and then remove it from any other modules that import the `shared` module. While you may believe that the `shared` module is unnecessary for small applications, it will save you a lot of repetitive import lines as your application grows larger, thus it is a smart practice to follow.

Creating components and services

Following that, let's develop the modules' components and services, beginning with `header`, `footer`, `page-not-found`, and `search-dialog` components that may be part of the root application module:

```
ng generate component core/components/header --module=app
```
```
ng generate component core/components/footer --module=app
```
```
ng generate component core/components/page-not-found
--module=app
```
```
ng generate component core/components/search-dialog
--module=app
```
```
ng generate service core/services/auth/auth
```

> **Important Note**
>
> Alternatively, we may create a `core` module that contains these components and import and re-export modules such as `BrowserModule`, `AppRoutingModule`, and `BrowserAnimationsModule`, while leaving the app module empty. However, we've opted to utilize a `core/` folder rather than a distinct module to aggregate our application's main features and import them straight from the application module.

Following that, we generate the `users` module's components using the following commands:

```
ng generate component users/components/login   --module=users
```
```
ng generate component users/components/signup   --module=users
```
```
ng generate component users/components/profile   --module=users
```
```
ng generate service users/services/profile/profile
```

Following that, we'll create the `feed` module's component:

```
ng generate component feed/components/posts --module=feed
```

The command has automatically added the created components to their respective modules.

Following that, we create the `shared` module's components and service:

```
ng generate component shared/components/create-post
--module=shared
```
```
ng generate component shared/components/post --module=shared
```

```
ng generate component shared/components/dialog --module=shared
ng generate service shared/services/post/post --module=shared
```

Then, in the `src/app/feed/feed.module.ts` file, import the `shared` module as follows:

```
import { NgModule } from '@angular/core';
import { CommonModule } from '@angular/common';

import { FeedRoutingModule } from './feed-routing.module';
import { PostsComponent } from './components/posts/posts.
component';
import { SharedModule } from '../shared/shared.module';

@NgModule({
  declarations: [PostsComponent],
  imports: [
    CommonModule,
    FeedRoutingModule,
    SharedModule
  ]
})
export class FeedModule { }
```

Furthermore, you must import the `shared` module in the `app` and `users` modules, since it will be used to import components (such as Angular Material components) required by these modules.

That's all we need to do now to build the structure of our application with modules, components, and services. Some other artifacts will also be added in the following chapters of this book. During the next few chapters, we'll develop the underlying functionality of each component and service.

In the next section, we'll learn about Angular Router and how to use it to map the components of our application's user interface to specific URL routes and provide in-app navigation.

Understanding Angular routing

Often, you'll need to implement routing in your web applications because you can't fit all of your app's features on a single page — this creates a poor user experience and even makes some tasks impossible for users, such as, in our case, the ability to access individual profiles via their unique URLs or to make profiles findable by search engines via **Search Engine Optimization (SEO)**.

As an all-in-one platform, Angular has a built-in router module that simplifies the implementation of single-page apps with numerous views and navigation. Additionally, the Angular CLI may aid you by automatically configuring the routing module in your app during the project creation process.

Angular Router is a static routing library. In contrast to dynamic routing, the routes are specified at the application's initialization. You may implement routing in a single routing module that contains all routes for your application, or you can add routing to each module individually.

Following that, we'll explore how Angular Router works and how the Angular CLI automates the process of configuring routing in our project.

How the Angular Router works

Instead of making requests to the server to obtain a new page, you can manage what users see in your Angular application by displaying or hiding the views controlled by components. This is accomplished by **client-side routing** rather than server-side routing.

To provide a great user experience, your application should present a distinct view for each activity or related group of activities, based on your business domain, rather than incorporating all functions into a single view. Additionally, it should enable navigation between distinct views — this may be accomplished through the use of Angular Router and browser URLs.

Angular Router is a client-side JavaScript solution that operates by monitoring changes to the URL and mapping paths to views using a route/component mapping array.

You indicate how paths are mapped to components in your routing module, and the router handles displaying the appropriate component when a path is matched (that would be when a user navigates to a certain path that is associated with a routing entry).

The router also allows you to provide a general wildcard entry that will be used when no matching paths are identified.

If a component matches, the router will display it in `router outlet`. `router outlet` is an Angular directive that serves as a place for the component(s) that match the current URL.

> **Important Note**
>
> In Angular, a directive is a concept that allows you to expand HTML with custom tags that are only available in the context of your application. Many built-in directives, such as `NgFor` and `NgIf`, are provided by Angular, and you may create your own by using the `@Directive` decorator.

The router may be found in the `@angular/router` module. By importing this module, you will have access to all of the services, components, and directives required to build routing and navigation in your Angular project.

Returning to the section *Initializing the Angular project*, when you created your project, the Angular CLI asked if you wanted to add routing. By choosing yes, you told the CLI to automatically set up routing, resulting in the generation of the routing module within the `src/app/app-routing.module.ts` file. You also instructed the Angular CLI to generate a routing module for the `users` and `feed` modules, to provide routing for each module's components in the *Creating modules, services, and components* section using the `--routing` option.

To implement routing, all you need to do now is add the route/component mappings inside the routing modules, but first let's look at what the CLI has done to set up routing.

The <base href> tag

When you view the `src/index.html` file, you will notice that a `<base>` tag has been inserted as a child of the `<head>` tag. This assists the router in constructing navigation paths. This is an example of your Angular application's `index.html` file after you've added routing:

```
<!doctype html>
<html lang="en">
<head>
  <meta charset="utf-8">
  <title>Client</title>
  <base href="/">
  <meta name="viewport" content="width=device-width,
      initial-scale=1">
  <link rel="icon" type="image/x-icon" href="favicon.ico">
```

```
</head>
<body>
  <app-root></app-root>
</body>
</html>
```

The `<base href>` tag is an HTML tag that lets you define the base URL that will be used to determine the absolute URL for any relative URLs on your document.

The routing modules

The CLI has added a routing module to the `src/app/app-routing.module.ts` file for setting routing, as seen below:

```
import { NgModule } from '@angular/core';
import { Routes, RouterModule } from '@angular/router';

const routes: Routes = [];

@NgModule({
  imports: [RouterModule.forRoot(routes)],
  exports: [RouterModule]
})
export class AppRoutingModule { }
```

This code imports the `Routes` and `RouterModule` symbols from the `@angular/router` package that provides the necessary router APIs. Next, it defines a `routes` array of the `Routes` type. This array will be used to add any routes in our application.

Next, `RouterModule` is added via the `imports` array of the routing module, and its `forRoot()` method is called with the previously defined `routes` array. `RouterModule` is then exported from the routing module using the `exports` array.

`AppRoutingModule` is simply an Angular module that wraps an instance of `RouterModule` with the `routes` array that will be used to hold any routes in our application. `RouterModule` is a built-in module that provides the service and directives required to implement routing, such as `RouterLink`, `RouterLinkActive`, and `Router-Outlet`.

The same routing module was added to the `users` and `feed` modules.

The router outlet

A `router-outlet` directive is added to the root application component in addition to the routing module. Go to the `src/app/app.component.html` template and look for some HTML code containing a `<router-outlet>` custom tag, as seen here:

```
<!-- [...] -->
<router-outlet></router-outlet>
```

The router outlet is a directive imported with the other routing APIs from the `@angular/router` module, and it serves as a marker to tell the router where to insert and render the component(s) matching the browser's URL, based on the information passed to the routes array described earlier.

Importing the routing module

The CLI also imported `AppRoutingModule` in the root application module, which can be found in the `src/app/app.module.ts` file, using the following code:

```
import { NgModule } from '@angular/core';
import { BrowserModule } from '@angular/platform-browser';

import { AppRoutingModule } from './app-routing.module';
import { AppComponent } from './app.component';
import { HeaderComponent } from
  './core/components/header/header.component';
import { FooterComponent } from
  './core/components/footer/footer.component';
import { PageNotFoundComponent } from
  './core/components/page-not-found/page-not-
    found.component';

@NgModule({
  declarations: [
    AppComponent,
    HeaderComponent,
    FooterComponent,
    PageNotFoundComponent
  ],
  imports: [
    BrowserModule,
```

```
        AppRoutingModule
    ],
    providers: [],
    bootstrap: [AppComponent]
})
export class AppModule { }
```

AppRoutingModule is imported from the `app-routing.module.ts` file and added to AppModule `imports` array. The same was done in the `users` and `feed` modules for each routing module.

We've documented all of the steps that the Angular CLI took automatically to establish routing in our project. In the following part, we'll look at how to add routes to our previously created components in order to provide in-app navigation.

Understanding and adding routes

As previously stated, the Angular CLI automatically configured routing when creating our project, so all we need to do now is specify the routes to the components of our application that we previously created.

Not all components are routable – in other words, they should be mapped to particular paths. For example, the header component will be included in other components to render the header element of the page through its selector, therefore we do not need to include it in our routing module's `routes` array.

A route is a `Route` object that holds information about which component maps to which path. A path is a URL element that specifies which view should be navigated to.

A route may have one or more of the following attributes:

- `path`: A property that holds the route's path.
- `pathMatch`: A property for configuring the matching strategy that the router will use to match the path. It can have a `prefix` or a `full` value, with the `prefix` being the default.
- `component`: A property referring to the component that should be mapped to the route.
- `redirectTo`: A property that contains another path to which you wish to be redirected if the path in the `path` property matches.
- `loadChildren`: A property to lazy-load a module by the router.

These are the most commonly used features of a route, but we may also utilize other properties, which can be found in the official documentation.

Consider the following example: we want the router to render the `posts` component if we browse to the `/posts` URL. All we have to do is add the following route configuration to the `routes` array:

```
{ path: 'posts', component: PostsComponent }
```

We may also have special paths, such as the empty path, which refers to the root URL of your application's domain name, or a wildcard path marked with `**`, which refers to a default route that the router will match if the user navigates to a path that does not exist in the `routes` array. You may use it to display a 404 error page informing the user that the page does not exist, or you can simply redirect the visitor to another URL.

Following that, we'll learn about the router's matching strategies for matching routes.

Route-matching strategies

The matching algorithm in the Angular router provides two strategies: `prefix` and `full`, however you may also use custom matching strategies:

- The router checks if the start of the browser's URL is prefixed with the route's path and renders the appropriate component using the default `prefix` strategy.

- The router uses the `full` strategy to determine if the path segment of the browser's URL is identical to the path of the route.

If you need to match the entire path, just set the `pathMatch` attribute of the route to `full` strategy, as seen here:

```
{ path: 'posts', pathMatch: 'full', component:
   PostsComponent }
```

The scenario of an empty path is another typical example of matching the full path, because the default prefix strategy will match all paths since the empty string is part of any path. For example, we can use the following route configuration to redirect visitors to the `/home` path when they first visit our application:

```
{ path: '',  redirectTo: '/home, pathMatch: 'full' }
```

We employ `redirectTo` instead of the `component` attribute, which accepts the path to which users will be routed.

Now that we've learned about route-matching strategies and the distinction between `prefix` and `full` strategies, let's look at route parameters.

Route parameters

Routes can also be used to carry arguments. The Angular router supports dynamic paths and offers the APIs needed to access route parameters within matched components.

You may add a parameter to your route by putting a colon before the parameter's name, as seen here:

```
{path: 'post/:id' , component: PostComponent}
```

We made use of a parameter called `id`.

The route parameter's value may be obtained by using the `ActivatedRoute` service, which allows you to obtain information about a route coupled with a component loaded in a router outlet, as well as the `ParamMap` Observable, which provides a map of the query parameters.

> **Important Note**
> Consider an Observable to be a collection of data that may be available at some point in the future, not necessarily at the time it is defined. It's a different method of performing asynchronous operations in JavaScript that's more powerful than Promises. It's supported on Angular via the RxJS library and is tightly integrated into the framework, with several built-in APIs.

Adding routes

Let's add routes to the components we previously created in our application now that we've learned about routes, matching strategies, and parameters.

Return to the `src/app/app.component.html` file and delete all of the unnecessary HTML markup, leaving only the router outlet. Next, include the `header` and `footer` components respectively on top of and below the router outlet, as follows:

```
<app-header></app-header>
<main>
    <router-outlet></router-outlet>
</main>
<app-footer></app-footer>
```

Next, open the `src/app/users/users-routing.module.ts` file and start by adding these imports:

```
import { LoginComponent } from
   './components/login/login.component';
import { SignupComponent } from
   './components/signup/signup.component';
import { ProfileComponent } from
   './components/profile/profile.component';
```

Next, add a route to each component and redirect the empty route to the login route:

```
const routes: Routes = [
   { path: '' , pathMatch: 'full', redirectTo: 'login' },
   { path: 'login' , component: LoginComponent },
   { path: 'signup' , component: SignupComponent },
   { path: 'profile/:userId' , component: ProfileComponent }
];
```

We imported the components of the `users` module and added routes to them inside the `routes` array.

Next, open the `src/app/feed/feed-routing.module.ts` file and add this import:

```
import { PostsComponent } from
   './components/posts/posts.component';
```

Next, add a route to the `posts` component and redirect the empty route to the `posts` route:

```
const routes: Routes = [
   { path: '', pathMatch: 'full', redirectTo: 'posts' },
   { path: 'posts', component: PostsComponent }
];
```

We imported the component of the feed module and added a route for it inside the `routes` array.

Finally, open the `src/app/app-routing.module.ts` file and start by adding the following import:

```
import { PageNotFoundComponent } from
  './core/components/page-not-found/page-not-
    found.component';
```

Next, add the highlighted routes to the `routes` array:

```
const routes: Routes = [
  { path: '', pathMatch: 'full', redirectTo: 'feed' },
  { path: 'users', loadChildren: () =>
    import('./users/users.module').then(m => m.UsersModule)
      },
  { path: 'feed', loadChildren: () =>
    import('./feed/feed.module').then(m => m.FeedModule) },
  { path: '**', component: PageNotFoundComponent }
];
```

We imported the `PageNotFoundComponent` component and then added a wildcard route for this component. This will take users to this component if they try to visit a route that doesn't exist in our router configuration. We also used the `loadChildren` property to lazy-load the `users` and `feed` modules when users visit the `users` and `feed` paths.

You can learn more about `loadChildren` from `https://angular.io/guide/lazy-loading-ngmodules`.

Now that we have learned about routing in Angular and added routes to different modules and components in our application, let's proceed to add navigation links so users can navigate between different views of our app.

Adding navigation

Angular Router provides the `routerLink` and `routerLinkActive` directives, which have to be used instead of the regular `href` attribute in the `<a>` tag to create navigation links. `routerLinkActive` is used to mark an active link.

Open the `src/app/core/components/header.component.html` file and update it as follows:

```
<button routerLink="/feed/posts">Feed</button>
<button routerLink="/users/profile/ahmedbouchefra">My
```

```
  profile</button>
<button routerLink="/users/login">Login</button>
<button routerLink="/users/signup">Sign up</button>
```

We added four navigation buttons with the `routerLink` directive, which holds the target path that users will be taken to after clicking the corresponding button.

This is just for testing our navigation links; we'll change this component's template later to include buttons in more appropriate places. For example, we don't actually need to put the login and sign-up buttons in the header area because the user will be automatically redirected to the `login` component if they are not logged in.

Next, let's see how to get the `userId` parameter, passed to the `/profile` path, in the corresponding profile component. Open the `src/app/users/components/profile/profile.component.ts` file and update it as follows:

```
import { Component, OnInit } from '@angular/core';
import { ActivatedRoute } from '@angular/router';

@Component({
  selector: 'app-profile',
  templateUrl: './profile.component.html',
  styleUrls: ['./profile.component.css']
})
export class ProfileComponent implements OnInit {
  constructor(private route: ActivatedRoute) { }
  ngOnInit(): void {
    this.route.params.subscribe(params => {
      console.log(params['userId']);
    });
  }
}
```

We import and inject an instance of `ActivatedRoute` via the component constructor, and we subscribe to the `params` Observable to get the `userId` parameter. We'll see later how to use this information to query for the information and posts related to the currently logged-in user and render them on the profile page.

After implementing routing and navigation in our Angular frontend, let's introduce you to some tips for debugging your application.

Debugging Angular applications

Angular by default runs in development mode, which has debugging enabled and provides developers with the `enableProdMode()` method to enable the production mode. This also disables debugging and removes any debugging information from the final production bundles.

With the new Ivy renderer, we also have a set of APIs on the global `ng` object that you can use to invoke methods, update state, and access components right from your browser's console when your Angular application is running. Check out `https://angular.io/api/core/global` for more information.

After running your Angular application, open your browser's console; you should be able to see the `Angular is running in development mode. Call enableProdMode() to enable production mode` message:

Figure 6.2 – Angular is running in development mode message in the console

This simply tells you that your application is in development mode and has debugging enabled.

After finishing developing your application, you should turn off debugging by enabling the production mode using the `enableProdMode()` function, as follows:

```
import { enableProdMode } from '@angular/core';
enableProdMode();
```

Enabling production mode will tell Angular to stop doing many unnecessary things that are not required in production, which will both reduce the size of your final application bundles and increase the runtime performance.

In fact, if you open the `src/main.ts` file of your project, you should see that this method is already added:

```
import { enableProdMode } from '@angular/core';
import { platformBrowserDynamic } from '@angular/platform-
  browser-dynamic';

import { AppModule } from './app/app.module';
```

```
import { environment } from './environments/environment';

if (environment.production) {
  enableProdMode();
}
// [...]
```

This simply enables the production mode if the `environment.production` variable is set to `true`, which is the case in the `environments/environment.prod.ts` file used by the Angular CLI when you run the `ng build --prod` command.

Go ahead and select some of the DOM elements in your application, for example, the **Feed** button, and then right-click and click on **Inspect** (or *Ctrl + Shift + I* on Chrome) in the context menu.

Inside DevTools, we can reference the most recently selected DOM element using `$0`. Combining that with the `ng.getOwningComponent()` method, we can get the component that rendered that element in the console, as follows:

```
> ng.getOwningComponent($0)
< ▼HeaderComponent { __ngContext__: LComponentView_AppComponent(53)}
    ▶ __ngContext__: LComponentView AppComponent(53) [app-root, TView, 147, LRootView(31), nul...
    ▼ proto :
      ▶ constructor: class HeaderComponent
      ▶ ngOnInit: f ngOnInit()
      ▶ proto : Object
> |
```

Figure 6.3 – Getting the parent component of a DOM element using ng

Go back to the `src/app/core/components/header/header.component.ts` file and add the following method to the component:

```
invokeMe(){
  console.log("Invoked from the console");
}
```

Next, go back to the running application, select a button from the header area, run the `ng.getOwningComponent($0)` method, and assign the result to a component variable, as follows:

```
const component = ng.getOwningComponent($0)
```

You should be able to call the `invokeMe()` method, as follows:

```
component.invokeMe()
```

This is the result you'll see in the browser's console:

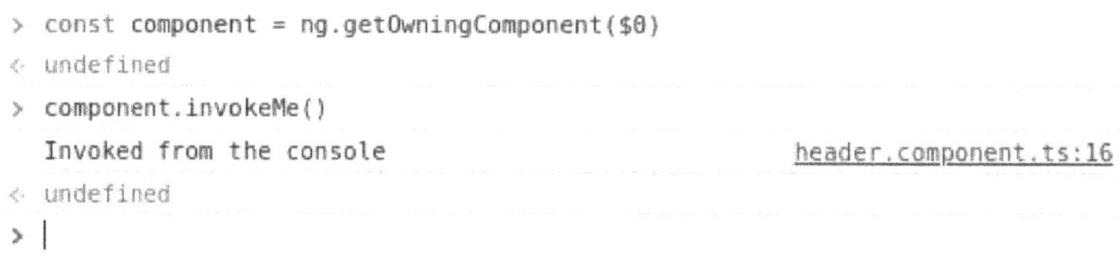

```
> const component = ng.getOwningComponent($0)
< undefined
> component.invokeMe()
  Invoked from the console                    header.component.ts:16
< undefined
> |
```

Figure 6.4 – Calling a component method from the console

You can also use other methods, such as the following:

- `ng.getComponent(element)` for getting the component of the selected element if it corresponds to a component

- `ng.getDirectives(element)` for getting the associated directives of the selected element

- `ng.applyChanges(element)` for triggering change detection of the selected component

We have seen that Angular runs by default in development mode with debugging enabled and how to turn on production mode when we finish developing our application. We've also seen some methods provided by Ivy that can help developers debug their Angular applications from the browser's console. In the next section, you'll learn about Angular Material, and you'll see how to install and configure it in the project using the Angular CLI.

Adding Angular Material

Angular Material provides UI components based on Material Design for Angular. Thanks to the Angular CLI, you can install it in your project in a few steps.

Head back to your command-line interface and run the following command from the root of your project:

```
ng add @angular/material
```

You'll be prompted with `The package @angular/material@12.2.8 will be installed and executed. Would you like to proceed? (Y/n)`. Type `Y` and press *Enter*.

This will install the package from npm. Next, the CLI will prompt you to `Choose a prebuilt theme name, or "custom" for a custom theme`. Let's pick `Indigo/Pink`. Next, it will ask `Set up global Angular Material typography styles?` Type y. Next, it will ask `Set up browser animations for Angular Material?` Type y.

That's it – you should have Angular Material installed and configured in your project.

Summary

Over the course of this chapter, we have learned about how to use the Angular CLI to generate a brand-new Angular application based on the latest version of the framework. We also looked at many essential concepts of Angular, such as modules, components, and services, and how to use the Angular CLI to generate those artifacts that provide the structure of our application.

After covering those topics and generating the basic structure of our application, we then discussed the basic concepts of routing in Angular, including the routing module, routes, and navigation directives such as `routerLink` that replace the regular `href` attribute in HTML.

Next, we looked at some tips on how to debug our application, using the debugging methods available by default in development mode, and why it's necessary to enable production mode when you are ready to deploy your application for production.

Finally, we added Angular Material to our project to provide Material Design components for building our application UI throughout the next chapters.

That concludes this chapter! In the next one, we'll continue our journey of building a frontend application to consume the GraphQL API exposed by the backend previously built. We'll specifically see how to integrate the frontend with the backend using Apollo Client for sending GraphQL queries and writing data on the server.

7
Adding User Search Functionality

In the previous chapter, we saw how to use the Angular CLI to create a fresh Angular project based on a recent version of the framework. We reviewed many fundamental Angular concepts, such as modules, components, and services, as well as how to use the Angular CLI to generate the artifacts that constitute the structure of our application.

We also included Angular Material in our project to provide Material Design components for creating a visually appealing application UI.

In this chapter, we'll look specifically at how to integrate the frontend with the backend using Apollo Client, which is designed for sending GraphQL queries and mutations to the server to fetch and write data.

We'll go over the concepts and steps needed to add authentication with Angular, Apollo Client, and GraphQL. Along with this, we will be developing the following features:

- A signup and login interface that allows users to create accounts and authenticate themselves
- A header bar with buttons for navigating the application, signing in, and logging out, as well as a search bar for searching for network users
- A footer that displays ngSocial (c) 2021

We will cover the following topics:

- Introducing and installing Apollo Client
- Importing Angular Material components
- Signing up and logging users in

Technical requirements

To successfully complete this chapter's steps, you must first work through the previous chapter of this book.

You can optionally install the following:

- Angular DevTools
- DevTools for Apollo Client
- Visual Studio Code

If you're using Visual Studio Code, it is strongly advised to install the GraphQL for Visual Studio Code extension with the `ext install GraphQL.vscode-graphql` command. You can also use the `ext install capaj.graphql-codegen-vscode` command to integrate Codegen with Visual Studio Code and have it run every time your query/mutation is saved.

You should also be acquainted with the following technologies:

- JavaScript/TypeScript
- HTML
- CSS
- RxJS Observables; check out `https://angular.io/guide/observables` and `https://rxjs.dev/guide/observable`
- Angular testing; check out `https://angular.io/guide/testing`

This chapter's source code can be found at `https://github.com/PacktPublishing/Full-Stack-App-Development-with-Angular-and-GraphQL/tree/main/Chapter07`.

Introducing and installing Apollo Client

In this section, we'll introduce you to Apollo Client, which helps you integrate your Angular frontend with your GraphQL API built on top of Apollo Server. Then, we'll see how to install and configure Apollo Client in our Angular project.

Apollo Client is a JavaScript state management library that allows developers to manage local and remote data using GraphQL. It can be used to retrieve, cache, and update application data while also automatically synchronizing data changes with your UI.

Let's install Apollo Client in our Angular project:

1. Return to your Terminal and execute the following command from your Angular project's root folder:

   ```
   ng add apollo-angular
   ```

   The apollo-angular package will be installed and executed. You will be asked the following question:

   ```
   Would you like to proceed?
   ```

   Type Y and press Enter.

2. Following that, you'll be prompted to enter the URL to your GraphQL endpoint. Type in http://localhost:8080/graphql.

   This will install and configure Apollo Client. An src/app/graphql.module.ts file is created and imported in the src/app/app.module.ts file, as follows:

   ```
   // [...]
   import { GraphQLModule } from './graphql.module';
   import { HttpClientModule } from
     '@angular/common/http';

   @NgModule({
     declarations: [
      // [...]
     ],
     imports: [
       // [...]
       GraphQLModule,
       HttpClientModule
     ],
   ```

```
    providers: [],
    bootstrap: [AppComponent]
})
export class AppModule { }
```

In the GraphQL module, we use the `HttpLink` service from `apollo-angular/http` to link the client to the GraphQL server running at `http://localhost:8080/graphql`. We also use `InMemoryCache` from `@apollo/client/core` to create a memory cache for storing local and remote data. The `HttpLink` service makes use of `HttpClient` to communicate with the server, so we also imported `HttpClientModule` from `@angular/common/http` in the application root module.

Apollo provides the Apollo Link interface, which allows you to manage the network layer where you can plug in links to control data flow, or simply how queries are sent, between the client and the GraphQL server. The network behavior of your client can be described as a chain of links that run sequentially.

`HttpLink` is one example of these links. By default, Apollo Client sends GraphQL operations to a remote server via HTTP using Apollo Link's `HttpLink`. This default link is created automatically, which covers our initial use cases, but later, we'll define more advanced links and specify their execution order to extend the default networking behavior.

Read `https://www.apollographql.com/docs/react/api/link/introduction/` for more information about Apollo Link and see a list of custom links developed by the community at `https://www.apollographql.com/docs/react/api/link/community-links/`.

The in-memory cache is where Apollo Client stores the data fetched from the GraphQL server. You can find more information at `https://apollo-angular.com/docs/caching/configuration/`.

Now that we've added Apollo Client to our application and established a connection to our GraphQL server, we'll see how to import the Angular Material components required for building the application's UI in the following section.

Importing Angular Material components

Each Angular Material component is contained within its own module. These components are typically used in multiple places throughout our application, which is why we must import and re-export them from the shared module to avoid importing them in each module:

1. Open the `shared/shared.module.ts` file and add the following imports:

```
import { MatToolbarModule } from
  '@angular/material/toolbar';
import { MatButtonModule } from
  '@angular/material/button';
import { MatIconModule } from
  '@angular/material/icon';
import { MatInputModule } from
  '@angular/material/input';
import { MatCardModule } from
  '@angular/material/card';
import { MatFormFieldModule } from
  '@angular/material/form-field';
import { MatTooltipModule } from
  '@angular/material/tooltip';
import { MatSnackBarModule } from
  '@angular/material/snack-bar';
import { MatDialogModule } from
  '@angular/material/dialog';
import { MatProgressBarModule } from
  '@angular/material/progress-bar';
import { MatBadgeModule } from
  '@angular/material/badge';
```

The names of the components' modules are self-explanatory; for example, `MatToolbarModule` contains the Material Toolbar component that will be used to build the header of our application, but you can also consult the official docs at `https://material.angular.io/` for more information about each available UI component.

2. Next, we need to add these components to the `imports` and `exports` arrays of the shared module, as follows.

First, add the imported modules inside the `matModules` array:

```
const matModules = [
    MatToolbarModule,
    MatButtonModule,
    MatIconModule,
    MatInputModule,
    MatCardModule,
    MatFormFieldModule,
    MatTooltipModule,
    MatSnackBarModule,
    MatDialogModule,
    MatProgressBarModule,
    MatBadgeModule
];
```

Next, add them to the `imports` and `exports` arrays of the shared module, as follows:

```
@NgModule({
    declarations: [],
    imports: [
        CommonModule, ReactiveFormsModule, ...matModules
    ],
    exports: [ReactiveFormsModule, ...matModules]
})
export class SharedModule { }
```

Since these Angular Material components need to be added to both the `imports` and `exports` arrays of the shared module, we added them to the `matModules` array once, then used the `Spread` operator to include them. Check out https://basarat.gitbook.io/typescript/future-javascript/spread-operator.

After we've imported and re-exported the necessary Material UI components, we'll look at how to add functionality for signing up and authenticating users, including how to use the UI components to build our registration and login interfaces.

Signing up and logging users in

In this section, we'll look at how to use Apollo Client and the Apollo library for Angular to add user authentication to our Angular frontend.

Defining GraphQL documents

Let's begin by defining the GraphQL documents (queries and mutations) required for user registration, signing in, and searching:

1. Create a `shared/constants/auth.ts` file (make sure to create the `constants/` folder inside the `shared/` folder) and begin by adding the following import and two string constants, containing the keys for the access token and authenticated user, which will be used to store the corresponding information in the user's browser's local storage:

    ```
    import { gql } from 'apollo-angular';
    export const ACCESS_TOKEN: string = 'accessToken';
    export const AUTH_USER: string = 'authUser';
    ```

2. Add and export the register mutation that is sent to the backend API for registering users:

    ```
    export const REGISTER_MUTATION = gql'
    mutation register($fullName: String!, $username:
      String!, $email: String!, $password: String!){
        register(fullName:$fullName, username:$username,
          email:$email, password:$password){
        token, user { id fullName bio email username image
          coverImage postsCount createdAt }
      }
    }
    ';
    ```

    We use the `gql` function to wrap query and mutation strings, so they get parsed to GraphQL documents. It's a template literal tag for passing queries, mutations, subscriptions, and fragments to Apollo Client. It returns an object of the `DocumentNode` type, which is an object representation of a GraphQL query string after getting parsed. See `https://git.io/JikfN`.

We used the `mutation` operation type since we need to call a mutation on the GraphQL server. We named it `register` and provided the necessary arguments using GraphQL variables (since we need to dynamically pass the full name, username, email, and password of the user to register an account for them).

In the body of our mutation, we simply asked for the `register` field, which we've previously added to the `Mutation` type in our GraphQL server schema, and because the corresponding resolver will return an object containing the authentication token and the user saved in the database, we specified a selection set to request the required fields.

Check out `https://graphql.org/learn/queries/` for more information about operation types.

3. Add and export the login mutation for authenticating users:

```
export const LOGIN_MUTATION = gql'
mutation signIn($email: String!, $password: String!){
  signIn(email:$email, password:$password){
    token, user { id fullName bio email username image
      coverImage postsCount createdAt }
  }
}
';
```

Again, we used the `gql` template literal to build the GraphQL document that will be sent to the server to authenticate users. We used the `mutation` type to instruct the server to call the `signIn` field on the top-level `Mutation` type and we used variables to pass dynamic arguments, such as the name and password, which are required to authenticate users.

Since the `signIn` mutation field refers to an object with a token string and a user object, we use a selection set to ask for all the required information on the client side, including a JWT and the authenticated user.

4. Create a `shared/constants/user.ts` file and import the `gql` tag, as follows:

```
import { gql } from 'apollo-angular';
```

Then, add and export the user query for getting a user by ID:

```
export const USER_QUERY = gql'
query getUser($userId: ID!){
  getUser(userId:$userId){
    id fullName bio email username image coverImage
      postsCount createdAt }
}
';
```

Here, we use the `query` operation type and we pass the user ID as an argument to ask the API server to return an existing user with that ID. The `Query` type on the server contains the `getUser` field that we are querying and because it returns a user object, we provide a selection set to ask for the fields that we are interested in, on the client side.

5. Add and export the following query to search for users:

```
export const SEARCH_USERS_QUERY = gql'
query searchUsers($searchQuery: String!){
  searchUsers(searchQuery:$searchQuery){
    id fullName bio username image
  }
}
';
```

Here, we use a `query` operation type to ask for the `searchUsers` field on the top-level query of the GraphQL schema, which returns an array of users with full names that contain the search term passed as an argument. We then specify a selection set for selecting the user fields that we are interested in, such as the user's ID, full name, bio, username, and image.

6. Finally, create a `shared/index.ts` file (barrel file) and export the defined constants, as follows:

```
export * from './constants/auth';
export * from './constants/user';
```

Following the definition of the queries and mutations that will be parsed and sent to the server via Apollo Client, we must define the types of responses/results that will be returned from the API server.

Defining types for user authentication

TypeScript is a strongly typed programming language. We'll use this to write type-safe code, which helps prevent unexpected errors and enables auto-completion from IDEs for increased productivity.

Let's take a look at how to create TypeScript types for strongly typing server responses. We'll start with the User model, which represents a single user and encapsulates the fields corresponding to the GraphQL API's user information:

1. Create a shared/models/user.model.ts file and add, then export, a TypeScript interface with the following fields that refer to the user's details and their types:

```
export interface User {
    id: string;
    fullName: string;
    bio: string;
    username: string;
    email: string;
    image: string;
    coverImage: string;
    postsCount: number;
    createdAt: string;
}
```

The preceding code block specifies the format of the user data returned by the GraphQL API.

2. Create a shared/types/auth.ts file and start by importing the User model:

```
import { User } from '../models/user.model';
```

Then, to strongly type the authentication state, define and export the following interface:

```
export interface AuthState {
    isLoggedIn: boolean;
    currentUser: User | null;
    accessToken: string | null;
};
```

Next, define and export the responses' types for the `register` and `signIn`
mutations:

```
export type AuthResponse = {
  token: string;
  user: User;
};
export type RegisterResponse = {
  register: AuthResponse;
};
export type LoginResponse = {
  signIn: AuthResponse;
};
```

3. Create a `shared/types/user.ts` file and start by adding the following imports:

```
import { Observable } from 'rxjs';
import { User } from '../models/user.model';
```

Next, define and export the following types to type the `getUser` and
`searchUsers` queries' responses:

```
export type UserResponse = {
  getUser: User;
};
export type UsersResponse = {
  searchUsers: User[];
};
export type SearchUsersResponse = {
  data: Observable<UsersResponse>;
  fetchMore: (users: User[]) => void
};
```

We imported the `Observable` class from RxJS and the `User` model we defined
earlier, then defined and exported a bunch of types for typing the responses
returned from the server. The types define the shape of the response's data returned
from the server for each operation. We'll use `SearchUsersResponse` to type
variables containing the search user's query response plus a method for fetching
more data used for pagination.

4. Finally, open the `shared/index.ts` file and add the following exports:

```
export { User } from './models/user.model';
export * from './types/user';
export * from './types/auth';
```

In the following section, we'll look at how to build services that encapsulate the logic for sending requests with queries and mutations for user authentication.

We added user search functionality to the authentication service because it will be global to our application and accessed before the profile service, which is only loaded when we lazy load the `user` module.

Defining Apollo services

After defining the GraphQL queries and mutations, as well as the types of responses that will be returned after the requests are sent to the server, we must now create a slew of services for registering, authenticating, and searching for users.

The Apollo Client library for Angular supports two methods of communicating with the server:

- Using the Apollo service, an Angular service that includes methods such as `query()` and `watchQuery()` to send queries, as well as `mutate()` and `subscribe()` to send mutations and subscriptions.

- Built-in `Query`, `Mutation`, and `Subscription` services.

  `Mutation`, `Query`, and `Subscription` are generic TypeScript classes that accept a generic type parameter between angle brackets after the class name. We specify the expected response type for each service and the `document` property for the mutation, query, or subscription that will be sent to the server. Check out `https://apollo-angular.com/docs/data/services`.

> **Note**
> For more information about the generic classes in TypeScript, check out `https://www.typescriptlang.org/docs/handbook/2/generics.html#generic-classes`.

We'll use both approaches to communicate with our backend throughout this book. More information can be found in the official documentation at `https://apollo-angular.com/docs/`.

Now, let's create the services for registering, authenticating, and searching for users:

1. Create the `core/services/auth/graphql/register.service.ts` file and add the following code:

```
import { Injectable } from '@angular/core';
import { Mutation } from 'apollo-angular';
import { RegisterResponse, REGISTER_MUTATION } from
  'src/app/shared';

@Injectable({
  providedIn: 'root',
})
export class RegisterGQL extends
  Mutation<RegisterResponse> {
    document = REGISTER_MUTATION;
}
```

2. Create the `core/services/auth/graphql/login.service.ts` file and add the following code:

```
import { Injectable } from '@angular/core';
import { Mutation } from 'apollo-angular';
import { LOGIN_MUTATION, LoginResponse } from
  'src/app/shared';

@Injectable({
  providedIn: 'root',
})
export class LoginGQL extends Mutation<LoginResponse> {
  document = LOGIN_MUTATION;
}
```

3. Create the `core/services/auth/graphql/getuser.service.ts` file and add the following code:

```
import { Injectable } from '@angular/core';
import { Query } from 'apollo-angular';
import { UserResponse, USER_QUERY } from
  'src/app/shared';
```

```
@Injectable({
  providedIn: 'root',
})
export class GetUserGQL extends Query<UserResponse> {
  document = USER_QUERY;
}
```

We import the necessary APIs, such as the `Injectable` decorator to decorate classes to define Angular services such as, the `Mutation` and `Query` services of `apollo-angular`, the constants that contain the queries and mutations that will be sent to the server, and the response types.

Next, we create our GraphQL services by extending the `Mutation` or `Query` services and setting the `document` property to the appropriate query or mutation.

The authentication service, which encapsulates the user's authentication functionality, will be implemented next.

Implementing the user authentication service

The authentication service is a part of the app module into which all previously defined Apollo services will be injected.

This service will contain the authentication logic for the user and provide methods for the user's components to register, sign in, log out, check the authentication state, get a specific user by ID, and finally, search for users.

Let's get started with the implementation of the authentication service by taking the following steps:

1. Open the `core/services/auth/auth.service.ts` file and start by adding these imports:

    ```
    import { Apollo } from 'apollo-angular';
    import { Observable, BehaviorSubject } from 'rxjs';
    import { map, tap } from 'rxjs/operators';
    ```

    Also, add the following imports from the shared module:

    ```
    import {
      User,
      RegisterResponse,
      LoginResponse,
    ```

```
  AuthState,
  UserResponse,
  UsersResponse,
  SearchUsersResponse,
  SEARCH_USERS_QUERY,
  ACCESS_TOKEN,
  AUTH_USER
} from 'src/app/shared';
```

Next, import the GraphQL services from their corresponding source files:

```
import { GetUserGQL } from
  './graphql/getuser.service';
import { LoginGQL } from './graphql/login.service';
import { RegisterGQL } from
  './graphql/register.service';
```

2. Define the following public properties and private attribute:

```
authState: Observable<AuthState>;
isLoggedInAsync: Observable<boolean>;
private readonly authSubject:
  BehaviorSubject<AuthState>;
```

We defined two Observables, `authState` and `isLoggedInAsync`, and an `authSubject` private attribute, which provides all the different ways for storing the authentication state of the current user. These Observables can be used to react to realtime changes to the authentication state.

3. Next, inject the Apollo service and the GraphQL services via the service constructor, as follows:

```
constructor(
  private apollo: Apollo,
  private registerGQL: RegisterGQL,
  private loginGQL: LoginGQL,
  private getUserGQL: GetUserGQL) {}
```

Inside the body of the constructor, add the following code:

```
const localToken = this.getLocalToken();
let isLoggedIn = false;
if (localToken) {
   isLoggedIn = this.tokenExists() &&
     !this.tokenExpired(localToken);
}
this.authSubject = new BehaviorSubject<AuthState>({
   isLoggedIn: isLoggedIn,
   currentUser: this.getLocalUser(),
   accessToken: localToken
});
this.authState = this.authSubject.asObservable();
this.isLoggedInAsync = this.authState.pipe(map(state
   => state.isLoggedIn));
```

Inside the constructor, we created the instance of `BehaviorSubject` with an initial state retrieved from the browser's local storage (used for persisting the authentication state after leaving the application).

After creating the authentication subject, we call the `asObservable()` method to take only the Observable part of the subject and assign it to the `authState` property. This can be used by the other components or services to watch for the authentication state since it's declared as a public property.

A subject is both an Observer and an Observable; but to watch the state from the other components, we only need the Observable part.

We also created the `isLoggedInAsync` property, which exposes only the `isLoggedIn` part of the authentication state using the `map()` operator on the `authState` Observable.

To get the access token, the current user, and the logged-in state stored on the local storage of the user's browser, we use the `getLocalToken()`, `getLocalUser()`, `tokenExists()`, and `tokenExpired()` methods of the service, which we'll define in the next steps.

If a token exists in `localStorage` and is not expired, we set the `isLoggedIn` variable to `true` and provide it as an initial value to the authentication subject along with the current user and access token.

We used `BehaviorSubject` for storing the authentication state.

`BehaviorSubject` is a type of subject that is able to store a current value and emits it to all new subscribers. It should always be initialized with an initial value. In our case, we initialize it with an object of the `AuthState` type that we've previously defined. It contains three properties, the `isLoggedIn` Boolean property, which indicates whether the user is logged in or not, the `currentUser` property, which contains the currently logged-in user, and the access token.

> **Note**
>
> RxJS Observables, subjects, and operators allow us to build a reactive UI that reacts to changes in real time. You can read more about RxJS and reactivity at `https://www.techiediaries.com/javascript-reactive-asynchronous-code-rxjs-6-angular-10/`.

4. Define the following properties below the service constructor to get the logged-in state of the users and the currently logged-in user:

```
get isLoggedIn(): boolean {
  return this.authSubject.getValue().isLoggedIn;
}
get authUser(): User | null {
  return this.authSubject.getValue().currentUser;
}
```

Here, we use the `getValue()` method of `BehaviorSubject` to get the value of the `isLoggedIn` and `currentUser` attributes of the authentication state. This method allows us to get the value stored in the behavior subject synchronously instead of subscribing to the Observable and getting the value asynchronously. This is possible because a behavior subject always contains a value at any point in time.

5. Define the following methods to get the token and store the current user in the local storage of the user's browser:

```
getLocalToken(): string | null {
  return localStorage.getItem(ACCESS_TOKEN);
}
storeUser(user: User): void {
  localStorage.setItem(AUTH_USER,
    JSON.stringify(user));
}
```

6. Define the following private methods to get the stored user from the local storage
    and store the token:

```
private getLocalUser(): User | null {
  return JSON.parse(localStorage.getItem(AUTH_USER) as
    string) ?? null;
}
private storeToken(token: string): void {
  localStorage.setItem(ACCESS_TOKEN, token);
}
```

7. Define the following private methods to check whether a token exists in the local
    storage and whether it's expired or not:

```
private tokenExists(): boolean {
  return !!localStorage.getItem(ACCESS_TOKEN);
}
private tokenExpired(token: string): boolean {
  const tokenObj =
    JSON.parse(atob(token.split('.')[1]));
  return Date.now() > (tokenObj.exp * 1000);
}
```

8. Define the following private methods to update and reset the authentication state:

```
private updateAuthState(token: string, user: User) {
  this.storeToken(token);
  this.storeUser(user);
  this.authSubject.next({
    isLoggedIn: true,
    currentUser: user,
    accessToken: token
  });
}
private resetAuthState() {
  this.authSubject.next({
    isLoggedIn: false,
    currentUser: null,
    accessToken: null
```

```
    });
}
```

The first method stores the provided token and user arguments in the local storage, and then calls the `next()` method of the subject to publish the authentication state to all subscribers. This method will be called by the `register()` and `login()` methods that will be defined next.

The second method calls the `next()` method of the authentication subject to reset the authentication state. This will be called by the `logOut()`, `register()`, and `login()` methods of the service that will be defined next. It sets the `isLoggedIn` attribute to `false` and `currentUser` and `accessToken` to `null`.

9. Define the `register()` method for registering users, as follows:

```
register(
  fullName: string,
  username: string,
  email: string,
  password: string): Observable<RegisterResponse |
    null | undefined> {}
```

Inside the body of the method, add the following code:

```
return this.registerGQL
  .mutate({
    fullName: fullName,
    username: username,
    email: email,
    password: password
  }).pipe(map(result => result.data), tap());
```

The method calls the `mutate()` method of the injected instance of the `RegisterGQL` service to register a user with the provided information.

Next, inside the `tap()` method, add the following Observer object:

```
{
  next: (data: RegisterResponse | null | undefined) => {
    if (data?.register.token && data?.register.user) {
      const token: string = data?.register.token,
        user: User = data?.register.user;
      this.updateAuthState(token, user);
```

```
      }
    },
    error: err => {
      console.error(err);
      this.resetAuthState();
    }
  }
```

In the tap() method, we check whether we get a token and user from the server and we call the updateAuthState() method to publish the authentication state using the behavior subject. In case of errors, we call resetAuthState() to reset the authentication state.

The tap() operator is used to perform side effects. Check out https://rxjs. dev/api/operators/tap.

10. Define the login() method to authenticate users, as follows:

```
login(
  email: string,
  password: string): Observable<LoginResponse | null |
    undefined> {}
```

Next, inside the body of the method, add the following code:

```
return this.loginGQL
  .mutate({
    email: email,
    password: password
  }).pipe(map(result => result.data), tap());
```

Next, inside tap(), add the following observer object:

```
{
  next: (data: LoginResponse | null | undefined) => {
    if (data?.signIn.token && data?.signIn.user) {
      const token: string = data?.signIn.token, user =
        data?.signIn.user;
      this.updateAuthState(token, user);
    }
  },
```

```
    error: err => {
      console.error(err);
      this.resetAuthState();
    }
  }
```

In both the `login()` and `register()` methods, we call the `mutate()` method of the corresponding Apollo service to send the requests to the server. They, respectively, return Observables of the `LoginResponse` and `RegisterResponse` types.

We use the `tap()` operator to execute a side effect that is storing the token and the authenticated user on the local storage and publishing the authentication state to subscribers. The `mutate()` method takes an object that contains the parameters that will be passed to the mutation.

We use the `map()` operator to get the data object of the result instead of the whole object, which corresponds to our responses' types or could also possibly be `null` or undefined, so we need to add those types; otherwise, the TypeScript compiler will throw errors.

In the case of errors, we log the error to the console and call the `resetAuthState()` method to reset the authentication state.

11. Implement the `getUser()` method to fetch a user by ID from our GraphQL API, as follows:

```
getUser(userId: string): Observable<UserResponse> {
  return this.getUserGQL.watch({
    userId: userId
  }).valueChanges.pipe(map(result => result.data));
}
```

In this method, we use the `watch()` method of the `GetUserGQL` service, which extends the Apollo `Query` service, to send the query and keep watching it; we use the `map()` operator to get the data object of the result, instead of the full Apollo query result object, which corresponds to an object of the `Observable<UserResponse>` type.

As long as we didn't unsubscribe from the returned Observable, this query will stay active during the life of our application.

12. Implement the `searchUsers()` method, as follows:

```
searchUsers(searchQuery: string, offset: number, limit:
number): SearchUsersResponse {
  // [...]
  return { data:
    feedQuery.valueChanges.pipe(map(result =>
      result.data)), fetchMore: fetchMore };
}
```

In the body of the method, add the following code to send the query:

```
const feedQuery = this.apollo.watchQuery<UsersResponse>({
  query: SEARCH_USERS_QUERY,
  variables: {
    searchQuery: searchQuery,
    offset: offset,
    limit: limit
  },
  fetchPolicy: 'network-only',
});
```

For a given query, you can specify different fetch policies. Include `fetchPolicy` to accomplish this. In this case, we instruct Apollo Client to ignore the cache and retrieve data directly from the server by using the `network-only` fetch policy. As a result, we will never reuse locally cached data and will always send a network request to retrieve the query, ignoring any locally cached data.

A fetch policy specifies how the Apollo Client will use the cache for a certain query when it is executed. The `cache-first` policy is the default, which means that Apollo Client first checks the cache to see whether the result is there before sending a network request to get the result from the server. If the result is present, there is no need to send a network request.

By configuring the fetch policy for this query to be `network-only`, we ensure that Apollo Client always asks our server to retrieve the most up-to-date list of users that correspond to the search term entered by the user.

You can find more details about the supported fetch policies at `https://www.apollographql.com/docs/react/data/queries/#supported-fetch-policies`.

13. Then, add the following function, which wraps the `fetchMore()` method of Apollo Client:

```
const fetchMore: (users: User[]) => void = (users:
    User[]) => {
      feedQuery.fetchMore({
        variables: {
          offset: users.length,
        }
      });
    }
```

In this method, we pass the GraphQL query for searching users to the `query` attribute in the `Apollo.watchQuery()` method. The method returns a `QueryRef` object that contains the `valueChanges` property, which is an Observable. We return an object of the `SearchUsersReponse` type with a data property, which is an Observable of the `UsersResponse` type, and a `fetchMore` property, which is a function that can be used to fetch more users according to the passed pagination parameters.

> **Note**
>
> The `fetchMore()` method is provided by Apollo for data pagination. Check out `https://apollo-angular.com/docs/caching/interaction/#incremental-loading-fetchmore`.

The method takes a set of variables to be sent with the query. We also set the offset to `users.length` so that we can fetch users that aren't already fetched.

When obtaining paginated data, we'll be using offset-based pagination. As implemented in this approach, the paginated field receives two arguments: a `limit` argument, which determines the amount of data that we want to fetch, and an `offset` argument, which tells us where in the set of data the server should start when returning results for a certain query.

You can learn about pagination with Apollo Client at `https://apollo-angular.com/docs/data/pagination`.

The `watchQuery()` method returns an object of the `QueryRef` type, which contains the same methods as an RxJS Observable, but to get data, we need to subscribe to its `valueChanges` property, which exposes a standard RxJS Observable.

14. Define the following method for logging users out:

```
logOut(): void {
  localStorage.removeItem(ACCESS_TOKEN);
  localStorage.removeItem(AUTH_USER);
  this.resetAuthState();
}
```

We log out by removing the access token and the authenticated user from the local storage and resetting the authentication state.

15. Create the `src/app/core/index.ts` file and export the authentication service, as follows:

```
export { AuthService } from
  './services/auth/auth.service';
```

Following the implementation of the required service methods for handling authentication, we'll look at how to build and style the signup UI.

Building and styling the signup UI

In this section, we'll build and style the signup UI, which includes a form that users must fill out and submit in order to register an account.

This is a screenshot of what we'll be building:

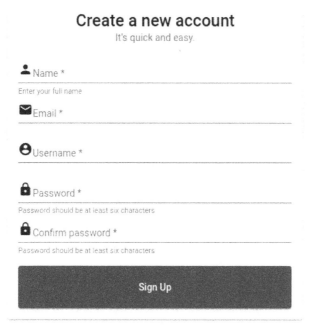

Figure 7.1 – The Signup UI

These are the steps you should take:

1. Open the `users/components/signup/signup.component.ts` file and start by adding the following imports:

    ```
    import { Component, OnInit, OnDestroy } from
       '@angular/core';
    import { FormGroup, FormControl, Validators } from
       '@angular/forms';
    import { Subscription } from 'rxjs';
    import { Router } from '@angular/router';
    import { MatSnackBar } from '@angular/material/
       snack-bar';
    import { AuthService } from 'src/app/core';
    import { RegisterResponse } from 'src/app/shared';
    ```

2. Implement the `OnDestroy` interface:

    ```
    export class SignupComponent implements OnInit,
       OnDestroy {}
    ```

3. Add a `form` property to the component:

```
form: FormGroup = new FormGroup({
  fullName: new FormControl('', [ Validators.required
    ]),
  email: new FormControl('', [ Validators.required,
    Validators.email]),
  username: new FormControl('', [ Validators.required
    ]),
  password: new FormControl('', [ Validators.required,
    Validators.minLength(6)]),
  password2: new FormControl('', [
    Validators.required, Validators.minLength(6) ])
});
```

Here, we create a reactive registration form, using `FormGroup`, `FormControl`, and `Validators`. We make sure all fields are required using `Validators.required`; we also ensure users enter a valid email address for the email field using `Validators.email` and a password with a minimal length of six characters using `Validators.minLength(6)`.

4. Add the `errorMessage` property and the private `registerSubscription` attribute:

```
errorMessage: string | null = '';
private registerSubscription: Subscription | null =
  null;
```

We define an `errorMessage` variable to hold any error message and display it on the template, and a `registerSubscription` variable that will hold the subscription returned from subscribing to the `register` method of the authentication service so we can unsubscribe from the returned `Observable` when the component is destroyed.

5. Inject the `MatSnackBar` router and `AuthService` into the component via the constructor:

```
constructor(
  private authService: AuthService,
  private snackBar: MatSnackBar,
  private router: Router) { }
```

6. Define the `ngOnDestroy()` method of the component, as follows:

```
ngOnDestroy() {
    if (this.registerSubscription) {
        this.registerSubscription.unsubscribe();
    }
}
```

In the `ngOnDestroy()` method, which gets called when the component is destroyed, we unsubscribe from the Observable returned from the `register()` method.

7. Define the `submit()` method, which gets called when the registration form is submitted:

```
submit() {}
```

Next, inside the method, destructure the `form.value` object:

```
const { fullName, username, email, password, password2
    } = this.form.value;
```

Next, check whether the password and its confirmation are the same; otherwise, return from the method:

```
if (password !== password2) {
    this.errorMessage = 'Passwords mismatch';
    return;
}
```

Then, check whether the form is valid; otherwise, return from the method:

```
if (!this.form.valid) {
    this.errorMessage = 'Please enter valid
        information';
    return;
}
```

Next, call the `register()` method and subscribe to the returned Observable:

```
this.registerSubscription = this.authService.
register(fullName, username, email,
    password).subscribe();
```

Finally, pass the following observer object to the `subscribe()` method:

```
{
    next: (result: RegisterResponse | null | undefined)
      => {
      const savedUserId = result?.register.user.id;
      this.snackBar.open('Signup Success!', 'Ok', {
        duration: 5 * 1000
      });
      if (savedUserId) {
      this.router.navigateByUrl
        ('/users/profile/${savedUserId}')
      }
    },
}
```

In the `submit()` method, we check whether the password and password confirmation are the same and whether the form is valid.

If the form is not valid, we return from the method and set the `errorMessage` variable to a convenient error message. Next, we call the `register()` method of the injected authentication service, and then we subscribe to the returned `Observable`.

In the next handler of the observer object passed to the `subscribe()` method, which will be called if the subscription is successful, we retrieve the registered user ID from the registration response, display a success message using a snack-bar, then navigate to the user's profile using the `navigateByUrl()` method of the router.

8. Add an error handler to the observer object, as follows:

```
error: (err) => {
    console.error(err.error);
    this.errorMessage = err.message;
    this.snackBar.open(err.message, 'Ok', {
      duration: 5 * 1000
    });
}
```

If there is an error, we display the error using a snack-bar and assign the error message to the `errorMessage` variable so it can also be displayed in the template.

9. Next, let's add a Material card that contains the HTML form; open the `src/app/` `users/signup/signup.component.html` file and update it, as follows:

```
<mat-card>
  <mat-card-title>Create a new account</mat-card-
    title>
  <mat-card-subtitle>It's quick and easy.</mat-card-
    subtitle>
  <mat-card-content>
    <!-- ADD_FORM_HERE -->
  </mat-card-content>
</mat-card>
```

10. Add the first section of the form, as follows:

```
<form [formGroup]="form" (ngSubmit)="submit()">
  <mat-form-field>
    <mat-icon matPrefix>person</mat-icon>
    <input autofocus type="text" matInput
      placeholder="Name *" formControlName="fullName">
    <mat-hint align="start">
      Enter your full name.
    </mat-hint>
  </mat-form-field>
```

We use the `formGroup` property to bind our form to the `form` group we created in the component's class, and we also bind the `ngSubmit` event of the form to the `submit()` method, which will be invoked when we submit the form. We use the `<mat-form-field>` component to wrap the form's fields to provide common Material styles and behavior to form controls such as the floating label, underlines, and hint messages. We use the `<mat-icon>` component with the `matPrefix` directive to display an icon in front of the form's field.

Next, we use the native `<input>` element with the `matInput` directive to allow the element to work with `<mat-form-field>`. We use the `autofocus` attribute to automatically give focus to the input element and `formControlName` to bind the element to the `fullName` property of the corresponding form's group defined in the component's class.

Finally, we add a hint message that appears below the underline, telling the user to enter the full name in this field, using the `<mat-hint>` component with the `align` attribute set to `start` to make it aligned to the left.

The form is not closed yet; we still need to add more fields in the next steps.

11. Add more fields to the form, as follows:

```
<mat-form-field>
  <mat-icon matPrefix>mail</mat-icon>
  <input type="email" matInput placeholder="Email *"
    formControlName="email">
</mat-form-field>
<mat-form-field>
  <mat-icon matPrefix>account_circle</mat-icon>
  <input type="text" matInput placeholder="Username
    *" formControlName="username">
</mat-form-field>
```

We create two form fields and bind them to the email and username attributes of the form's group in the component's class.

12. Add more fields to the form, as follows:

```
<mat-form-field>
  <mat-icon matPrefix>lock</mat-icon>
  <input type="password" matInput
    placeholder="Password *"
      formControlName="password">
  <mat-hint align="start">
    Password should be at least six characters.
  </mat-hint>
</mat-form-field>
<mat-form-field>
  <mat-icon matPrefix>lock</mat-icon>
  <input type="password" matInput
    placeholder="Confirm password *"
      formControlName="password2">
  <mat-hint align="start">
```

```
     Password should be at least six characters.
    </mat-hint>
  </mat-form-field>
```

We add two form fields for the password and its confirmation. We add a lock icon for both fields prefixed to the password input field; we use a <mat-hint> component to provide users with a hint message, saying that the password should be at least six characters long, placed on the left. We also add placeholder messages that display a hint to the user of what we expect to be entered in the fields.

13. Add the following markup and close the form, as follows:

```
<p *ngIf="errorMessage" class="error-message">
  {{ errorMessage }}
</p>
<button type="submit" mat-raised-button
  color="primary">Sign Up</button>
</form>
```

We conditionally render an error message if one exists using the ngIf directive and the errorMessage property defined in the component's class. Next, we display a submit button for the form, enhanced with Material Design styling and ink ripples using the mat-raised-button directive; then, we close the form.

14. After that, we'll need to add some CSS styles. First, proceed to the src/app/app.component.css file and update it as follows:

```
:host {
  min-height: 100%;
  display: flex;
  flex-direction: column;
  align-items: stretch;
}
main {
  flex-grow: 1;
}
```

We use the :host pseudo-class to style the element that hosts the component. We turn it into a flex container with a column direction so the elements inside the container will be displayed in columns and be stretched.

15. Then, open the `signup/signup.component.css` file and start by adding the following styles for the component's host element:

```css
:host {
    display: flex;
    justify-content: center;
    margin: 100px 0px;
}
```

We set the display to `flex`, justify the contents to the center, and finally, set the top and bottom margins to 100 pixels and the left and right margins to 0 pixels.

16. Style the card containing the form, as follows:

```css
mat-card-title,
mat-card-subtitle,
mat-card-content {
    display: flex;
    justify-content: center;
}
```

We set the display to `flex` and center the contents of the card's title, subtitle, and body.

17. Style the form fields, the button, and the error message, as follows:

```css
.mat-form-field {
    width: 100%;
    min-width: 400px;
    display: block;
}
.mat-raised-button {
    width: 100%;
    height: 60px;
    margin-top: 15px;
}
.error-message {
    color: white;
    background-color: red;
}
```

We set the width of the form fields to 100% and the minimal width to 400 pixels; we set the display to `block`. Next, we set the width of the button to 100%, the height to 60 pixels, and the top margin to 15 pixels. Finally, we set the color of the error message to white and the background color to red.

Following the creation and styling of the signup component, we will proceed to the creation and styling of the login component.

Building and styling the login UI (assignment)

In this assignment, build and style the login UI, which should include a form that users must fill out and submit in order to sign in. The form should look like this:

Figure 7.2 – The login UI

These are the steps you should take:

1. Begin by adding the necessary imports to the `login.component.ts` file. Next, implement the `OnDestroy` interface, which includes the `ngOnDestroy` method, and then create a group for the login form, which includes two `email` and `password` controls that must be bound to the corresponding HTML form's fields in the associated template. Both an email address and a password are required, and the password should be at least six characters long. The `required`, `email`, and `minLength` validators should be used to enforce these rules.

2. Define an `errorMessage` property of the `string` type that should contain any errors, as well as a private attribute for holding the subscription created by subscribing to the Observable returned by the authentication service's `login()` method.

3. Next, inject `AuthService`, `MatSnackBar`, and the router via the component's constructor, then unsubscribe from the login Observable in the component's `ngOnDestroy` life cycle event, which will be called when the component is destroyed and removed from the DOM, to avoid any memory issues.

4. Implement the `submit()` method. In the method's body, first destructure the form's value object to get the email and password values entered by the user; then check whether the form is valid. If it's not valid, add an error message and exit the method.

5. Then, using the provided email and password, call the authentication service's `login()` method and subscribe to the returned Observable to send the request to the server.

    You must get the user ID from the response object returned by the server in the `next` handler of the observer object passed to the `subscribe()` method, then display a **Login Success!** message and redirect the user to their profile page. If there is an error, you should display it using a snack-bar.

6. Insert the login form into the component's template, which can be found in the `login.component.html` file. The HTML form must be placed within a Material card. Bind its `formGroup` property to the component's `form` property, then bind its `ngSubmit` event to the component's `submit()` method, which will be called when you submit the form.

7. Using the `<mat-form-field>` element, add an email field and bind it to the `email` attribute of the login form's group in the component's class. The `matInput` directive is then applied to the field to add Material Design styling and behavior.

8. Add a password field and bind it to the login form's `password` attribute with a hint message on the left stating that the password should be at least six characters, using the `<mat-hint>` element.

9. Using the `ngIf` directive and the `errorMessage` property of the component's class, conditionally render the error message if it exists. Then, if the user does not already have an account, a link will be displayed to direct them to the registration page.

10. Add the login form's submit button and style it with Material Design, using the `mat-raised-button` directive, and a primary color.

11. In the `login.component.css` file, add the following CSS styles to style the login UI.

    Set the display to `flex`, justify the contents to the center, and set the top and bottom margins to 100 pixels and the left and right margins to 0 pixels to style the component's host element.

The form's fields should be styled with a block display and a width of 100%.

Set the display to `flex` and justify the contents to the center of the card containing the form.

The error message should have a width of 300 pixels, a white-colored text, and a red background.

The submit button should have a full width of 300 pixels, a height of 60 pixels, and a top margin of 15 pixels.

That's it; if you completed the assignment successfully, you should see a login UI similar to the one shown in *Figure 7.2*. This assignment's source code can be found at this link: `https://git.io/Ja7Bo`.

Implementing the header component

Let's implement and style the header and footer components now that we've implemented the user authentication functionality.

When the user is logged in, the header should look like this:

Figure 7.3 – The header component

Begin with the header component and proceed as follows:

1. Open the `core/components/header/header.component.ts` file and start by adding the following imports:

```
import { Component, OnInit, ElementRef, ViewChild }
  from '@angular/core';
import { Router } from '@angular/router';
import { MatDialog, MatDialogConfig } from
  '@angular/material/dialog';
import { User, SearchUsersResponse } from
  'src/app/shared';
import { AuthService } from 'src/app/core';
import { SearchDialogComponent } from '../search-
  dialog/search-dialog.component';
```

2. Add the following two public properties to the header component:

```
@ViewChild('searchInput') searchInput: ElementRef |
  null = null;
fetchMore: (users: User[]) => void = (users: User[])
  => { };
```

To access the corresponding input element, referenced by the searchInput reference in the template, where users can enter the term for searching for other users in our application, we define a searchInput property of the ElementRef type and decorate it with @ViewChild.

The fetchMore property is then defined, which accepts a function for retrieving more paginated users.

Then, via the component's constructor, inject AuthService, MatDialog, and Router:

```
constructor(
    public authService: AuthService,
    public matDialog: MatDialog,
    private router: Router) { }
```

3. Implement the logOut() method in the header component, as follows:

```
logOut(): void {
    this.authService.logOut();
    this.router.navigateByUrl('/users/login');
}
```

We call the logOut() method of AuthService, then navigate to the login page.

Then, implement the method to search for users:

```
searchUsers(): void {
    const searchText =
      this.searchInput?.nativeElement.value;
    const result: SearchUsersResponse =
      this.authService.searchUsers(searchText, 0, 10);
    const dialogConfig = new MatDialogConfig();
    dialogConfig.data = result;
    this.matDialog.open(SearchDialogComponent,
```

```
    dialogConfig);
}
```

We get the search term entered by the user in the search input field using the
nativeElement.value property of the searchInput property, then call the
searchUsers() method of the authentication service to send the search query to
the server, and finally, call the openSearchDialog() method to display the users
whose names correspond to the entered search term, using a Material dialog.

4. Open the core/components/header/header.component.html file and
    add the following markup to display the search bar on the left:

```html
<mat-toolbar>
    <mat-toolbar-row>
        <h1>ngSocial</h1>
        <div *ngIf="authService.isLoggedIn">
            <input class="searchbar" type="text"
                placeholder="Search" #searchInput matInput>
            <button mat-icon-button
                (click)="searchUsers()">
                <mat-icon>search</mat-icon>
            </button>
        </div>
        <span class="spacer"></span>
```

We add a Material toolbar using the <mat-toolbar> component with one
row using the <mat-toolbar-row> component. Next, we add the <h1> tag
containing our application's name, as well as a <div> tag containing the search
bar and some navigation buttons that will be added next.

We conditionally render the <div> tag containing the input field, which has
the #searchInput template reference used to access it from the component's
class with the ViewChild decorator. It also has a search icon button for invoking
the searchUsers() method defined in the component's class and bound to the
click event.

If the user is logged in, the search bar will be displayed. Using the mat-icon-
button and matInput directives, we apply Material Design styling and behavior
to the button and input field. The <mat-icon> component is used to display icons.

We also add a span with a spacer class to add space between the input field on the
left and the buttons on the right of the toolbar.

5. Add the following markup to display the buttons to show the notifications and navigate to the user's profile:

```
<button *ngIf="authService.isLoggedIn" mat-icon-button
  matTooltip="Notifications">
    <mat-icon aria-hidden="false" aria-
      label="Notifications">
        notifications</mat-icon>
</button>
<button *ngIf="authService.isLoggedIn" mat-icon-button
  matTooltip="My profile"
    routerLink="/users/profile/{{authService.authUser?.
id}
  }">
    <mat-icon aria-hidden="false" aria-label="My
      profile">person</mat-icon>
</button>
```

If the user is logged in, we conditionally render the button to display notifications and the button to navigate to the user's profile. We obtain the current user's ID by accessing the `id` property of the authentication service's `authUser` property, which is injected publicly via the component's constructor.

Because the buttons only contain icons and no text, we use the `matTooltip` attribute of the **Material** button to display a tooltip or hint when hovering over them, informing users about their function.

6. Add the following markup to display the buttons to navigate to the feed and login pages, as well as the logout button:

```
<button *ngIf="authService.isLoggedIn" mat-icon-button
  matTooltip="Home" routerLink="/feed/posts">
    <mat-icon aria-hidden="false" aria-
      label="Home">home</mat-icon>
</button>
<button *ngIf="!authService.isLoggedIn" mat-icon-
button matTooltip="Log In" routerLink="/users/login">
    <mat-icon aria-hidden="false" aria-label="Log
```

```
        in">login</mat-icon>
  </button>
  <button *ngIf="authService.isLoggedIn" mat-icon-button
    matTooltip="Log Out" (click)="logOut()">
      <mat-icon aria-hidden="false" aria-label=
        "Log out">logout</mat-icon>
  </button>
```

Finally, close the toolbar:

```
    </mat-toolbar-row>
  </mat-toolbar>
```

If the user is logged in, we conditionally render the button for navigating to the home feed and the logout buttons. If the user is not logged in, we render the login button.

Then, we close the toolbar row and the toolbar component. For the home and login buttons, we use the router's `routerLink` directive to navigate to the appropriate routes, while for the logout button, we simply bind the `logOut()` method defined in the component's class to the logout button's click event.

7. Open the `core/components/header/header.component.css` file and add the following CSS styles:

```
.spacer {
  flex: 1 1 auto;
}
button.mat-raised-button {
  border-radius: 0;
}
.searchbar {
  width: 99%;
  margin: 5px;
  height: 30px;
  border-radius: 23px;
}
```

Implementing the footer and page not found components

After we've finished with the header component, we'll move on to the footer component and then the page not found component.

This is how our footer appears:

ngSocial (c) 2021

Figure 7.4 – The footer component

Let's start with the steps:

1. Open the `core/components/footer/footer.component.html` file and update it as follows:

    ```
    <p>
       ngSocial (c) 2021
    </p>
    ```

2. Open the `core/components/footer/footer.component.css` file and add the following style:

    ```
    p {
       margin: 1rem;
    }
    ```

    We set the margin of the p element in the footer component to `1rem`.

    Let's now add the page not found component.

    If you go to any path that does not exist in our router configuration after implementing this component, you should see a page that looks like this:

This page doesn't exist!

Figure 7.5 – The page not found component

3. Open the core/components/page-not-found/page-not-found.
   component.html file and update it as follows:

```
<p>
  This page doesn't exist!
</p>
```

4. Open the core/components/page-not-found/page-not-found.
   component.css file and update it as follows:

```
:host {
  display: flex;
  justify-content: center;
  margin: 100px 0px;
}
```

We style the component's host element with the flex display to justify the contents to the center, and then we set the margin.

We've now implemented and styled our header, footer, and page not found components; now let's look at how to implement the user search dialog component.

Implementing the user search dialog component

The user can use the search bar in the header to find other users who have created accounts in our application. A dialog component will be created and displayed after entering the search term and clicking the **Search** button.

In this section, we'll look at how to make the search dialog component shown in the following screenshot:

Figure 7.6 – The search dialog component

1. Go to the `search-dialog/search-dialog.component.ts` file in the `core/components/` folder and start by adding the following imports:

```
import {
  Component,
  Inject,
  OnInit,
  OnDestroy
} from '@angular/core';
import { Router } from '@angular/router';
import { Subscription } from 'rxjs';
import {
  MatDialogRef,
  MAT_DIALOG_DATA
} from '@angular/material/dialog';
import {
  User,
  SearchUsersResponse,
  UsersResponse   } from 'src/app/shared';
```

2. Add the following properties to the component's class:

```
usersSubscription: Subscription | null = null;
users: User[] = [];
fetchMore: (feed: User[]) => void = () => {};
```

Next, update the constructor, as follows:

```
constructor(
  public dialogRef:
    MatDialogRef<SearchDialogComponent>,
  @Inject(MAT_DIALOG_DATA) public response:
    SearchUsersResponse,
  public router: Router) { }
```

3. Update the ngOnInit() method, as follows:

```
ngOnInit(): void {
  this.usersSubscription =
    this.response.data.subscribe({
    next: (data: UsersResponse | null | undefined) => {
      if (data?.searchUsers) {
        this.users = data.searchUsers;
      }
    }
  });
  this.fetchMore = this.response.fetchMore;
}
```

4. Add the ngOnDestroy() method, as follows:

```
ngOnDestroy(): void {
  if (this.usersSubscription) {
    this.usersSubscription.unsubscribe();
  }
}
```

5. Add the following helper methods to the component, which will be called from the template:

```
goToProfile(profileId: string): void {
    this.router.navigateByUrl('/users/profile/${profileId}
    ');
}
invokeFetchMore(): void {
    this.fetchMore(this.users);
}
close(): void {
    this.dialogRef.close();
}
```

6. Next, open the search-dialog/search-dialog.component.html file in the core/components/ folder and add the following contents:

```
<h2 mat-dialog-title>Users</h2>
<mat-dialog-content>
<!- […] -->
</mat-dialog-content>
<mat-dialog-actions>
<button class="mat-raised-button"
    (click)="close()">Close</button>
<button class="mat-raised-button"
    (click)="invokeFetchMore()">More..</button>
</mat-dialog-actions>
```

Then, within the dialog's content section, iterate through the users and display each one with a Material card:

```
<mat-card *ngFor="let user of users;">
    <mat-card-header>
        <img mat-card-avatar [src]="user.image?? ''" />
        <mat-card-title>
            {{ user.fullName }}
        </mat-card-title>
        <mat-card-subtitle>
            {{ user.username }}
            <button (click)="goToProfile(user.id)"
```

```
      mat-button>Visit</button>
    </mat-card-subtitle>
  </mat-card-header>
</mat-card>
```

We've completed our signup and login interfaces, which will allow users to register and sign in. Then, we built the header and footer components. The header contains navigation buttons for our application that are rendered conditionally based on whether the user is authenticated or not. We only show the button that takes the user to the login component if the user is not logged in.

The header also includes a search bar, which allows users to search for other users in our social network and display the results on a Material Design-styled dialog component.

Summary

Throughout this chapter, we've seen how to use Apollo Client to send queries and mutations to our previously implemented backend API. We integrated Apollo with our Angular frontend, then implemented the required services and components for user authentication.

We have also added functionality to our application header to allow users to navigate between pages, such as the profile, home feed, and login pages. We used conditional rendering to display just the necessary buttons based on the user's authentication state. We added the necessary functionality to allow users to search for other users on the network using a search bar in the header. The results are then displayed in a dialog component that includes pagination and close buttons.

In the next chapter, we'll continue improving our authentication system and add unit testing to ensure our code works as expected.

8
Guarding Routes and Testing Authentication

In the previous chapter, we saw how to use Apollo Client to send queries and mutations to our previously implemented backend **API**. We integrated Apollo with our Angular frontend, then implemented the required services and components for user **authentication**.

We also added functionality to our application header to allow users to navigate between pages such as the profile, home feed, and login pages. We added the necessary functionality to allow users to search for other users on the network using a search bar in the header.

In this chapter, we will continue implementing our auth system by guarding the necessary route(s) against unauthorized access, sending the **JWT** with the API requests, and unit testing our code.

We will cover the following topics in this chapter:

- Guarding routes
- Testing the auth service and component(s)

Technical requirements

To successfully complete this chapter's steps, you must first finish the previous chapter of this book.

You should also be acquainted with the following technologies:

- JavaScript/TypeScript.
- **HyperText Markup Language (HTML)**.
- **Cascading Style Sheets (CSS)**.
- Angular testing—check out `https://angular.io/guide/testing`.

This chapter's source code can be found at `https://github.com/PacktPublishing/Full-Stack-App-Development-with-Angular-and-GraphQL/tree/main/Chapter08`.

Guarding routes

Since we've already implemented auth, we'll need to protect the application's routes from non-authenticated users.

Not all pages should be protected; for example, the user's profile page could be made public so that users who do not yet have an account can find other users via search engines. If they have friends on the network, this will encourage them to sign up for an account.

Let's begin by creating an auth guard, as follows:

1. Return to your Terminal and type the following command:

    ```
    ng g guard core/guards/auth/auth
    ```

    You'll be prompted with `Which interfaces would you like to implement?`.

    Press *spacebar* to select `CanActivate`, then *Enter* to confirm.

2. Open the `core/guards/auth/auth.guard.ts` file and start by importing the auth service and `Router`, like this:

    ```
    import { ActivatedRouteSnapshot, CanActivate, Router,
    RouterStateSnapshot, UrlTree } from '@angular/router';
    import { AuthService } from
    '../../services/auth/auth.service';
    ```

3. Inject the auth service and `Router` via the constructor of the guard's class, as follows:

```
constructor(
    private authService: AuthService,
    private router: Router){}
```

4. Next, update the `canActivate()` method of the auth guard, as follows:

```
canActivate(
    route: ActivatedRouteSnapshot,
    state: RouterStateSnapshot): Observable<boolean |
    UrlTree> | Promise<boolean | UrlTree> | boolean |
        UrlTree {
            If (this.authService.isLoggedIn) {
              return true;
            }
            else {
              return this.router.parseUrl
                ("/users/login");
            }
        }
```

In the guard's `canActivate()` method, which will be called to determine whether the route can be activated or not, we use the `isLoggedIn` property of the injected auth service to determine whether the current user is logged in; if so, we return `true`. Otherwise, we use the router's `parseUrl()` method to redirect the user to the login component. This method returns an `UrlTree` object, which is nothing more than a parsed **Uniform Resource Locator (URL)**.

We tell Angular to redirect users to the login component by returning an URL tree structure of the `/users/login` URL.

> **Note**
>
> For more information about `UrlTree`, check out `https://angular.io/api/router/UrlTree`.

5. Export the auth guard from the `index.ts` file of the `core/` folder, as follows:

```
export { AuthGuard } from './guards/auth/auth.guard';
```

6. Open the `src/app/app-routing.module.ts` file and import the guard:

```
import { AuthGuard } from 'src/app/core';
```

Next, update the feed route in the `routes` array, as follows:

```
{
    path: 'feed',
    canActivate: [AuthGuard],
    loadChildren: () =>
        import('./feed/feed.module').then(m =>
            m.FeedModule)
}
```

To apply the auth guard we just created, we use the route's `canActivate` option. This protects the `feed` module's child routes from unauthenticated users.

We've added auth to our app and secured the feed routes from public access. Users of our application can now create an account, sign in, and search for other network users using the search input in the header area (actually, this last functionality is not fully implemented yet).

Because we now have two apps, a frontend and backend, we can run the following command from the root of our **monorepo** project to start both:

```
lerna run start --stream
```

This will run the `start` script of the client and server packages.

Sending the JWT

If you try to search for users at this point, you will receive an error stating that the user is not authenticated. This is due to the fact that the access token must be sent along with the request.

Here's how you can do this:

1. Open the `src/app/graphql.module.ts` file and start by adding the following highlighted imports:

```
import { ApolloClientOptions, from, InMemoryCache }
    from '@apollo/client/core';
import { HttpLink } from 'apollo-angular/http';
import { setContext } from
```

```
'@apollo/client/link/context';
import { HttpHeaders } from '@angular/common/http';
```

2. Update the createApollo() function, as follows:

```
export function createApollo(httpLink: HttpLink ):
ApolloClientOptions<any> {
  const accessToken =
    localStorage.getItem('accessToken');
  const http = httpLink.create({ uri });
  const setAuthorizationLink = setContext(() => ({
    headers: new HttpHeaders().set(
      'Authorization', 'Bearer ${accessToken}'
    )
  }));
  return {
    link: from([setAuthorizationLink, http]),
    cache: new InMemoryCache(),
  };
}
```

For more details, check out https://www.apollographql.com/docs/
react/api/link/apollo-link-context/.

Improving the search functionality (assignment)

As an assignment, update the code of the header component's searchUsers() method,
as follows:

- If the user enters an empty search term, display a message saying **Please enter a
search term!** using the Material snack bar (import and inject MatSnackBar), after
which the method should be exited.

- Check whether any users are found before opening the dialog to show the results;
otherwise, display the **No users found with this search term!** message using a
snack bar.

This assignment's source code can be found at this link: https://git.io/Ja7WN.

Using GraphQL fragments

In this section, we'll refactor our code to take advantage of another GraphQL feature: **GraphQL fragments**.

Multiple queries and mutations make use of the same user's fields that we've used in multiple GraphQL documents. Using fragments, we can avoid repeating these fields.

Begin by refactoring our queries to make use of GraphQL fragments. A fragment allows you to define a set of fields that you can reuse across multiple queries without having to repeat the same ones.

To refactor existing queries to use fragments, follow these steps:

1. Create a `shared/constants/user.fragments.ts` file and add the following contents:

    ```
    import { gql } from 'apollo-angular';


    export const BASIC_USER_FIELDS_FRAGMENT = gql'
    fragment BasicUserFields on User {
      id fullName bio username image
    }';
    export const USER_FIELDS_FRAGMENT = gql'
    fragment UserFields on User {
      id fullName bio email username image coverImage
        postsCount createdAt
    }';
    ```

2. Next, let's update our query strings to use the previous fragments in their selection sets in order to reduce repeated fields that are shared by queries. Open the `shared/constants/auth.ts` file and begin by importing the following fragments:

    ```
    import { USER_FIELDS_FRAGMENT } from
        './user.fragments';
    ```

3. Then, update the `LOGIN_MUTATION` document, as follows:

    ```
    export const LOGIN_MUTATION = gql'
    ${USER_FIELDS_FRAGMENT}
    mutation signIn($email: String!, $password: String!){
      signIn(email:$email, password:$password){
    ```

```
      token, user { ...UserFields }
    }
  }
';
```

4. Finally, update the `REGISTER_MUTATION` document, as follows:

```
export const REGISTER_MUTATION = gql'
${USER_FIELDS_FRAGMENT}
mutation register($fullName: String!, $username:
  String!, $email: String!, $password: String!){
  register(fullName:$fullName, username:$username,
    email:$email, password:$password){
      token, user { ...UserFields }
    }
  }
';
```

As an assignment, update the documents in the `shared/constants/user.ts` file using the `UserFields` and `BasicUserFields` fragments.

Local state management with Apollo Client

In this section, we'll see how to refactor our code to make use of Apollo Client's built-in local state management features, such as **local-only fields**, **field policies**, and **reactive variables**.

Reactive variables, introduced in Apollo Client 3, are a convenient way to store local state outside of the Apollo Client cache. They can hold information of any type and structure because they are independent of the cache, and you can interact with them from anywhere in your application without using GraphQL syntax.

> **Note**
>
> Read `https://www.apollographql.com/docs/react/local-state/reactive-variables/` for more explanations about reactive variables.

Local-only fields are simply those that are not defined in the schema of the GraphQL server. These values are computed locally using whatever logic you want, such as reading data from `localStorage`.

> **Note**
>
> Read https://www.apollographql.com/docs/react/
> local-state/managing-state-with-field-policies/.

A field policy defines the custom logic that determines how a single GraphQL field is fetched from and written to the Apollo Client cache. Field policies can be set for both local-only and remotely fetched fields.

> **Note**
>
> Please read https://www.apollographql.com/docs/react/
> caching/cache-field-behavior/ before proceeding.

We used an RxJS behavior subject to manage the auth state in our auth service, which is a common approach in Angular, but because we're using Apollo Client, which is also a state management tool, we can take advantage of its cache, as well as features such as local-only fields and reactive variables, to manage the auth state, making the cache the only **source of truth** in our app.

Let's take it one step at a time to see how we can refactor our code, particularly the auth service and the header component, to make use of Apollo Client cache and reactive variables, as follows:

1. Create a new `src/app/cache.ts` file and move the code for creating an in-memory cache there, like this:

    ```
    import { InMemoryCache } from '@apollo/client/core';
    export default new InMemoryCache();
    ```

2. Go back to the `src/app/graphql.module.ts` file and import the cache, as follows:

    ```
    import cache from './cache';
    ```

    Then, modify the `createApollo()` function to use the imported cache, like this:

    ```
    export function createApollo(httpLink: HttpLink ):
      ApolloClientOptions<any> {
      // [...]
      return {
        link: from([setAuthorizationLink, http]),
    ```

```
    cache: cache
  };
}
```

Also, make sure to remove the unused InMemoryCache import from this module.

3. Create an src/app/reactive.ts file and include the following imports:

```
import { makeVar } from '@apollo/client/cache';
import { gql } from '@apollo/client/core';
import { AuthState } from './shared';
```

Next, create and export a reactive variable named authState using the makeVar() function, as follows:

```
export const authState = makeVar<AuthState>(
  {
    accessToken: '',
    currentUser: null,
    isLoggedIn: false
  } as AuthState
);
```

This will initialize our reactive variable with the provided object, which should have the shape of the AuthState interface passed as a generic type between brackets.

The function returned by makeVar() can be used to get the variable's current value by calling authState() and to set a new value by calling authState(newValue).

The documentation for reactive variables can be found at https://apollo-angular.com/docs/local-state/reactive-variables/.

4. Next, define and export the following client-side query:

```
export const GET_AUTH_STATE = gql'
  query getAuthState {
    authState @client
  }
';
```

We create a GraphQL document (using the `gql` tag) with a query named `getAuthState` that asks Apollo Client for a field named `authState`. We use the `@client` directive to inform Apollo Client that this is a client-side query and that it should not attempt to send a request to the server. It will instead retrieve it from its cache.

5. Next, you need to use a field policy to add an `authState` field to the `Query` type policy in the cache. Go back to the `src/app/cache.ts` file and import the reactive variable, as follows:

```
import { authState } from './reactive';
```

Then, update the in-memory cache by adding a field policy for `authState` to the `Query` type policy, as follows:

```
export default new InMemoryCache({
    typePolicies: {
        Query: {
            fields: {
                authState: {
                    read() {
                        return authState();
                    }
                }
            }
        }
    }
});
```

When the `authState` field is queried, the `read()` function returns the value of the `authState` reactive variable. In our case, we've given the field and reactive variable the same name, but this isn't required.

To provide the logic that controls how to read from or write to the corresponding GraphQL field in the Apollo Client cache, we use a field policy. Field policies can be added for either local-only fields (with the `@client` directive appended) or remote fields (fetched from the server).

> **Note**
>
> For more information, check out `https://apollo-angular.com/docs/local-state/management#field-policies-and-local-only-fields`.

6. Open the `core/services/auth/auth.service.ts` file and start by adding these imports:

```
import { ApolloQueryResult } from
  '@apollo/client/core';
import { authState, GET_AUTH_STATE } from
  'src/app/reactive';
```

Then, from the service's class, remove the following members:

```
authState: Observable<AuthState>;
isLoggedInAsync: Observable<boolean>;
private readonly authSubject: BehaviorSubject<AuthState>;
```

Next, inside the constructor, remove the following code:

```
this.authSubject = new BehaviorSubject<AuthState>({
  isLoggedIn: isLoggedIn,
  currentUser: this.getLocalUser(),
  accessToken: localToken
});
this.authState = this.authSubject.asObservable();
this.isLoggedInAsync = this.authState.pipe(map(state
  => state.isLoggedIn));
```

Then, replace it with the following code:

```
authState({
  isLoggedIn: isLoggedIn,
  currentUser: this.getLocalUser(),
  accessToken: localToken
});
```

We remove the code that initializes `BehaviorSubject` and other Observables that hold the auth state and replace it with code that initializes our reactive variable with the initial auth state retrieved from the browser's `localStorage`.

You should also remove the import for the disused `BehaviorSubject` symbol.

7. To obtain the auth state, define the following property:

```
get authState(): Observable<AuthState> {
  return this.apollo.watchQuery<{ authState: AuthState
    }>({
```

```
        query: GET_AUTH_STATE
    }).valueChanges.pipe(map((qr: ApolloQueryResult<{
        authState: AuthState }>) => qr.data.authState));
}
```

Also, make the following changes to the existing isLoggedIn getter:

```
get isLoggedIn(): Observable<boolean> {
    return this.apollo.watchQuery<{ authState: AuthState
}>({
        query: GET_AUTH_STATE
    }).valueChanges.pipe(map((qr: ApolloQueryResult<{
        authState: AuthState }>) =>
        qr.data.authState.isLoggedIn));
}
```

Then, update the existing authUser getter, as follows:

```
get authUser(): Observable<User | null> {
    return this.apollo.watchQuery<{ authState: AuthState
}>({
        query: GET_AUTH_STATE
    }).valueChanges.pipe(map((qr: ApolloQueryResult<{
        authState: AuthState }>) =>
        qr.data.authState.currentUser));
}
```

In all of these getter properties, we invoke Apollo Client's watchQuery() method, passing the client-side query for getting the auth state as a parameter to send the query to the cache, which will respond with the result of the authState field. The client will use the field policy's read() function to return the result, which in our case is the value of the reactive variable that holds the auth state.

8. Update the private updateAuthState() method by removing the code for publishing the new state of the behavior subject and replace it with a call to update the reactive variable, as follows:

```
private updateAuthState(token: string, user: User) {
    this.storeToken(token);
    this.storeUser(user);
    authState({
        isLoggedIn: true,
```

```
    currentUser: user,
    accessToken: token
  });
}
```

Then, update the `resetAuthState()` method, as follows:

```
private resetAuthState() {
  authState({
    isLoggedIn: false,
    currentUser: null,
    accessToken: null
  });
}
```

To update the auth state, the `register()`, `login()`, and `logOut()` methods call these updated methods. Rather than calling the subject's `next()` method to publish the new auth state, we use the `authState()` function to update the reactive variable containing the local auth state.

That's it—we've finished refactoring our service; see the commit at `https://git.io/JVyRO`.

Now, we'll look at the auth guard and header component, as follows:

1. Go to the `core/guards/auth/auth.guard.ts` file and update the auth guard by first importing the `map()` operator, like this:

    ```
    import { map } from 'rxjs/operators';
    ```

    Then, in the auth guard's class, update the `canActivate()` method, as follows:

    ```
    canActivate(
      route: ActivatedRouteSnapshot,
      state: RouterStateSnapshot): Observable<boolean |
        UrlTree> | Promise<boolean | UrlTree> | boolean |
        UrlTree {
          return this.authService.isLoggedIn.pipe(
          map((isLoggedIn: boolean) => {
            if (isLoggedIn) {
              return true;
            }
            return this.router.parseUrl('/users/login');
    ```

```
      })
    );
  }
```

Instead of the synchronous (current) value returned by the behavior subject's getValue() method, our auth guard now uses the Observable, returned by the updated isLoggedIn property, and the router's parseUrl() method.

2. Then, you need to update the header component. Open the core/components/header/header.component.ts file and start by adding the necessary imports, as follows:

Update the first import statement from @angular/core to include the OnDestroy symbol, as follows:

```
import { Component, OnInit, ElementRef, ViewChild,
  OnDestroy } from '@angular/core';
```

Update the import from src/app/shared to include AuthState, as follows:

```
import { User, SearchUsersResponse, UsersResponse,
  AuthState } from 'src/app/shared';
```

Then, update the import statement from rxjs/operators to include the takeUntil symbol, like this:

```
import { first, takeUntil } from 'rxjs/operators';
```

Finally, import the Subject symbol, as follows:

```
import { Subject } from 'rxjs';
```

3. Implement the OnDestroy interface by executing the following code:

```
export class HeaderComponent implements OnInit,
  OnDestroy { /*...*/ }
```

Next, add the following members to the component:

```
public isLoggedIn: boolean = false;
public authUser: User | null = null;
private destroyNotifier$: Subject<boolean> = new
  Subject<boolean>();
```

To avoid memory leaks, we'll use the subject to notify the `takeUntil()` operator when the component is destroyed so that it can complete the source Observable it's applied to. Read the official documentation at `https://rxjs.dev/api/operators/takeUntil`.

4. Update the component's `ngOnInit()` method, as follows:

```
ngOnInit(): void {
  this.authService.authState
  .pipe(takeUntil(this.destroyNotifier$))
  .subscribe({
    next: (authState: AuthState) => {
      this.isLoggedIn = authState.isLoggedIn;
      this.authUser = authState.currentUser;
    }
  });
}
```

We subscribe to the Observable returned by the `authState` property and provide an observer object with a `next()` handler that is called whenever the Observable emits a value (which, in our case, is the auth state stored in the reactive variable).

To avoid memory leaks, we also use the `takeUntil()` operator to complete the Observable when the component is destroyed.

Then, add the `ngOnDestroy()` method, as follows:

```
ngOnDestroy(): void {
  this.destroyNotifier$.next(true);
  this.destroyNotifier$.complete();
}
```

When the component is destroyed, we notify the `takeUntil()` operator by invoking the subject's `next()` method, which emits a `true` value, and we also complete the notifier subject.

5. Go to the `core/components/header/header.component.html` file and change the `*ngIf` expressions from using the `authService.isLoggedIn` property, which used to return a Boolean value but now returns an Observable, to using the component's `isLoggedIn` property. For example, the search bar should be rendered as follows:

```
<div *ngIf="isLoggedIn">
    <input class="searchbar" type="text"
      placeholder="Search" #searchInput matInput>
    <button mat-icon-button (click)="searchUsers()">
        <mat-icon>search</mat-icon>
    </button>
</div>
```

You must do the same thing to conditionally render the buttons on the right.

Also, for the button that takes users to their profiles, replace the interpolated `authService.authUser?.id` expression between `{{` and `}}` with just `authUser?.id`, as follows:

```
<button *ngIf="isLoggedIn" mat-icon-button
    matTooltip="My profile"
    routerLink="/users/profile/{{authUser?.id}}">
```

For all updates made to the header component, check out this commit: `https://git.io/JVyW0`.

If you check Apollo Client Devtools after you log in, you should see an active `getAuthState` query that will remain active as long as the header component is not destroyed, as illustrated in the following screenshot:

Figure 8.1 – getAuthState query in Apollo Client Devtools

Using the OnPush change detection strategy

Angular employs the `ChangeDetectionStrategy.Default` strategy by default. This strategy is not the most efficient in terms of performance because every time an event occurs in our application, such as clicks or network responses, change detection runs on all components. To avoid unnecessary change detection runs, we can use the `OnPush` strategy for some components, as shown in the following steps:

1. Go to the `core/components/header/header.component.ts` file and import `ChangeDetectionStrategy` and `ChangeDetectorRef` from the `@angular/core` module, as follows:

    ```
    import { Component, OnInit, ElementRef, ViewChild,
      OnDestroy, ChangeDetectionStrategy,
        ChangeDetectorRef } from '@angular/core';
    ```

2. Change the detection strategy to `OnPush`, like this:

    ```
    @Component({
      selector: 'app-header',
      templateUrl: './header.component.html',
      styleUrls: ['./header.component.css'],
      changeDetection: ChangeDetectionStrategy.OnPush
    })
    export class HeaderComponent implements OnInit,
      OnDestroy {
    /* [...] */}
    ```

    Afterward, you will notice that the search bar and the buttons on the right are not properly rendered if you log out and log back in. This occurs because Angular does not run change detection to synchronize the view with the component's updated variables—namely, the `authUser` and `isLoggedIn` members. Let's fix that by manually triggering change detection whenever the auth state changes.

3. Inject `ChangeDetectorRef` into the header component's constructor, like this:

    ```
    constructor(
      /*...*/,
      private changeDetectorRef: ChangeDetectorRef) { }
    ```

4. Mark the component for checking in the header component's `ngOnInit()` method, as follows:

```
ngOnInit(): void {
  This.authService.authState
    .pipe(takeUntil(this.destroyNotifier$))
    .subscribe({
      next: (authState: AuthState) => {
        this.isLoggedIn = authState.isLoggedIn;
        this.authUser = authState.currentUser;
        this.changeDetectorRef.markForCheck();
      }
    });
}
```

Use the `OnPush` detection strategy for the footer and page-not-found components as an assignment. The updates and solutions can be found at this commit: `https://git.io/JV9zb`.

Testing the auth service and component(s)

Before we finish implementing the auth section, we need to add automatic tests to our implementation to detect any hidden errors that we might miss when manually testing.

When we generate our project with the **Angular command-line interface** (**Angular CLI**), we also have testing configured with a number of initial tests that are added to ensure that artifacts, such as components and services, are properly instantiated.

If you're new to Angular testing, start by reading `https://angular.io/guide/testing` and `https://techiediaries.com/angular/jasmine-unit-testing`, then learn how to write your first test suite at `https://jasmine.github.io/tutorials/your_first_suite`.

Also, before testing Apollo Client APIs, make sure to read the official testing documentation at `https://apollo-angular.com/docs/development-and-testing/testing/`.

Jasmine includes a set of APIs that make it simple to write unit tests. We basically have three options, as follows:

- The `describe()` function is used to define a set of tests.

- The `it()` function is used to initialize a unit test/specification.

- The expect () function is used to define expectations to ensure that our code works as expected.

When we generate an artifact, such as a component or service, with the Angular CLI, we also get a specification file that includes tests for the artifact, as well as a unit test that checks whether the artifact is instantiated.

Testing the auth service

Your project is ready for testing. All you have to do is follow the upcoming steps in this section.

Go to your Terminal and type the following command:

```
ng test
```

The ng test command builds the application and runs the Karma test runner in watch mode. Visit https://github.com/karma-runner/karma for more information. This command will run our tests and open a web browser with either passed or failed tests. At this point, we would expect all tests to pass, but this is not the case. This is due to the fact that the tests, not the code, are broken.

This is because we modified the default project's code generated by the Angular CLI by injecting some services via the constructor or adding code to the components' ngOnInit () method. Because they are involved in the creation of components, the constructor and ngOnInit () methods are both called in the existing unit tests.

Let's start by testing the auth service before we fix the broken tests. If you are unfamiliar with testing services, read https://angular.io/guide/testing-services to better understand these steps.

Most of our service's methods follow a similar pattern—we request data, handle errors, and extract data from the response, so they should be tested in much the same way.

We must mock/fake the response and then assert that queries and mutations are called once or multiple times with the expected documents and variables, and that the mocked response data is returned. Proceed as follows:

1. Return to your Terminal and run the tests for the auth service only, using the following command from the client/ folder:

   ```
   ng test --include src/app/core/services/auth/auth.
   service.spec.ts
   ```

As you can see, the existing *should-be-created* test case passes; let's write some more. Open the `core/services/auth/auth.service.spec.ts` file and add the following specs:

```
it('should register a user');
it('should reset auth state when registration fails on
    server');
it('should authenticate a user');
it('should reset auth state when login fails on
    server');
it('should store user and token on updateAuthState
    call');
it('should call localStorage.setItem on storeUser
    call');
it('should call localStorage.setItem on storeToken
    call');
it('should return the isLoggedIn state');
it('should return the authenticated user state');
it('should return the full auth state');
it('should log out users');
it('should search for users');
it('should get user by ID');
```

If you go to your Karma report before implementing the actual test logic, in the second parameter of the `it()` function, you'll see that your specs are marked as pending, as illustrated in the following screenshot:

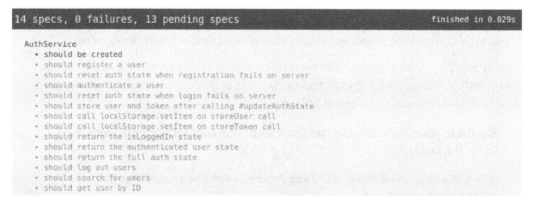

Figure 8.2 – Pending tests of the auth service

We'll use the `apollo-angular/testing` module, which includes `ApolloTestingModule` and `ApolloTestingController`, to make it simple to test code that calls Apollo Client APIs by mocking calls to the GraphQL endpoint.

2. Begin by including the following imports:

```
import {
  ApolloTestingController,
  ApolloTestingModule
} from 'apollo-angular/testing';
import {
  AuthResponse,
  RegisterResponse,
  REGISTER_MUTATION,
  User
} from 'src/app/shared';
```

3. Let's now extend the `describe()` block by defining a testing controller, as follows:

```
let controller: ApolloTestingController;
```

Then, create a fake user object, as follows:

```
const fakeUser: User = {
  id: 'id#1',
  fullName: 'A B',
  username: 'a.b',
  email: 'a.b@email.com'
} as User;
```

Create a fake token, as follows:

```
const fakeToken: string = '<ADD_A_TOKEN_HERE>';
```

You can get a valid token from your server using Apollo Studio, or use an online service such as `https://token.dev/`.

4. Then, using the fake user and token, create an auth response, as follows:

```
const authResponse: AuthResponse = {
  token: fakeToken,
  user: fakeUser
};
```

The testing module must then be configured by adding `ApolloTestingModule` to the `imports` array and injecting the testing controller, as shown in the following code snippet:

```
beforeEach(() => {
  TestBed.configureTestingModule({
    imports: [ApolloTestingModule]
  });
  service = TestBed.inject(AuthService);
  controller =
    TestBed.inject(ApolloTestingController);
});
```

5. After that, add the `afterEach()` function, which will be called after each test in the suite, and call the testing controller's `verify()` method, as follows:

```
afterEach(() => {
  controller.verify();
});
```

After configuring testing, let's implement the `should register a user` test.

6. Start by updating the `it()` function, as follows:

```
it('should register a user', (done) => {});
```

Next, inside the callback's block, define a fake register response, like this:

```
const fakeRegisterResponse: RegisterResponse = {
  register: authResponse
};
```

Then, call the `spyOn()` function to spy on the `updateAuthState()` method of the service, as follows:

```
spyOn(service, 'updateAuthState' as never);
```

Since this method is private, we need to add `as never` to stop TypeScript from throwing errors.

7. Next, call the `register()` method of the service, as follows:

```
service.register('A B', 'a.b', 'a.b@techiediaries.com',
'1..9').subscribe({
  next: (result) => {
```

```
    expect(result)
      .toEqual(fakeRegisterResponse);
    expect(service['updateAuthState'])
      .toHaveBeenCalledOnceWith(
        fakeRegisterResponse.register.token,
        fakeRegisterResponse.register.user
      );
    done();
  }
});
```

We call the method with the same user information that was added to the fake response and then subscribe to the returned Observable. In the observer object's `next()` handler, we assert that the returned result equals the fake response and that the `updateAuthState()` method was called with the correct arguments. Finally, we use the `done()` function to notify Jasmine that our asynchronous task has been completed.

We used introspection to access the method because `updateAuthState()` is private.

8. Outside of the `register()` method, we must verify that one request containing the register mutation was sent by using the following code:

```
const op = controller.expectOne((operation) => {
  expect(operation.query.definitions)
    .toEqual(REGISTER_MUTATION.definitions);
  return true;
});
```

9. Assert that the operation variables have the same values as the `register()` method's passed parameters by executing the following code:

```
expect(op.operation.variables.fullName)
  .toEqual('A B');
expect(op.operation.variables.username)
  .toEqual('a.b');
expect(op.operation.variables.email)
  .toEqual('a.b@techiediaries.com');
expect(op.operation.variables.password)
  .toEqual('1..9');
```

10. Finally, flush the fake response to make the Observable emit, as follows:

```
op.flush({ data: fakeRegisterResponse });
```

You should have two passed specs after saving and returning to your Karma report, as indicated in the following screenshot:

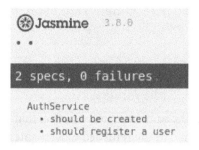

Figure 8.3 – Passed tests

After we've tested the register method, let's add a second spec to assert that the auth state is reset when registration fails on the server.

11. Begin by including the following import:

```
import { authState }
  from 'src/app/reactive';
```

Then, update the corresponding it() function, as follows:

```
it('should reset auth state when registration fails on
  server ', (done) => {});
```

Add the following object, which holds the initial auth state:

```
const initialState = {
  isLoggedIn: false,
  currentUser: null,
  accessToken: null
};
```

12. Spy on the private resetAuthState() method, as follows:

```
const resetAuthStateSpy =
  spyOn(service, 'resetAuthState' as never);
```

Call the register() method to assert that our auth state is equal to the initial state and that the observer object's error() handler has called the resetAuthState() method, like this:

```
service.register('A B', 'a.b',
  'a.b@techiediaries.com', '1..9').subscribe({
  error: (err) => {
    expect(authState())
      .toEqual(initialState);
    expect(resetAuthStateSpy)
      .toHaveBeenCalled();
    done();
  }
})
```

We obtained the current auth state using the authState() function and then compared it to the initial state.

13. Assert that the correct operation was used by executing the following code:

```
const op = controller.expectOne((operation) => {
  expect(operation.query.definitions).toEqual
    (REGISTER_MUTATION.definitions);
  return true;
});
```

Finally, use the following method to cause the Observable to emit a network error:

```
op.networkError({} as Error);
```

As an assignment, implement the following test cases:

1. should authenticate a user
2. should reset auth state when login fails on server
3. should store user and token on updateAuthState call
4. should call localStorage.setItem on storeUser call
5. should call localStorage.setItem on storeToken call

Check out the implementation at this commit: https://git.io/Johq3.

Then, for the `isLoggedIn` property, let's add a test case. Unfortunately, it is unclear how to use the controller to mock the response for the client-side query without causing errors, so we will implement this spec by spying on Apollo Client's `watchQuery()` method, as shown next:

1. Begin by adding the necessary imports, as follows:

    ```
    import { GET_AUTH_STATE } from 'src/app/reactive';
    import { Apollo, QueryRef } from 'apollo-angular';
    import { of } from 'rxjs';
    ```

2. Then, define the following `Apollo` object:

    ```
    let apollo: Apollo;
    ```

3. Inject the Apollo service as follows in the `beforeEach()` function:

    ```
    apollo = TestBed.inject(Apollo);
    ```

    Then, update the test case, as follows:

    ```
    it('should return the isLoggedIn state', (done) =>
        {});
    ```

4. Inside the body, spy on the `apollo` object's `watchQuery()` method and return a fake `QueryRef` object, as follows:

    ```
    spyOn(apollo, 'watchQuery').and.returnValue({
      valueChanges: of(
        {
          data:
          {
            authState:
            {
              isLoggedIn: true
            }
          }
        })
    } as QueryRef<any, any>);
    ```

Then, subscribe to the isLoggedIn getter and assert that the watchQuery()
method has been called with the appropriate query and that the isLoggedIn
state is true, as follows:

```
service.isLoggedIn.subscribe((isLoggedIn: boolean) => {
  expect(apollo.watchQuery)
    .toHaveBeenCalledOnceWith({ query: GET_AUTH_STATE });
  expect(isLoggedIn).toEqual(true);
  done();
});
```

As an assignment, implement the following specifications in the same manner:

- should return the authenticated user state
- should return the full auth state

Check out the implementation at this link: https://git.io/Johtr.

Implement the following specifications as assignments as well:

- should log out users
- should search for users
- should get user by ID

Please note that you need to add the __typename field to the getUser
and searchUsers queries in the shared/constants/user.ts file for
ApolloTestingController to work properly with queries and fragments.

Check out the implementation at this link: https://git.io/JohLH.

During the testing of our auth service, we had the opportunity to think deeply about our
implementation so that we could test it correctly, which also aids in detecting hidden
errors and improving code quality. For example, resetting the auth state after a failed
registration or login isn't really necessary, so let's change our code, as follows:

1. Remove error handling from the auth service's register and login methods.
2. To avoid repeating null | undefined in the service's code, use a generic
   type alias. Make sure to include this alias in the login and registration components
   as well.
3. Using the exit() function, disable the related tests.

Check out this link for updates: https://git.io/JohPJ.

Testing the auth guard

Let's write the auth guard tests. We'll write some tests to assert that the guard redirects users appropriately based on their auth status. If users are already logged in, our guard uses the auth service to determine whether they are already logged in; otherwise, it redirects them to the login path using the router's parseUrl() method.

To ensure that our implementation works as expected, we need to add an isolated unit test suite. As a result, we must define a stub of the auth service that the auth guard will use to determine whether or not the user is authenticated. We'll also define a spy object for the parseUrl() method. Follow these steps:

1. Open the core/guards/auth/auth.guard.spec.ts file and begin by configuring the tests by including the following imports:

    ```
    import { TestBed }
      from '@angular/core/testing';
    import {
      ActivatedRouteSnapshot,
      Router,
      RouterStateSnapshot,
      UrlTree } from '@angular/router';
    import { Observable, of } from 'rxjs';
    import { AuthService }
      from '../../services/auth/auth.service';
    ```

    Then, add the following class for mocking the auth service:

    ```
    class AuthServiceMock {
      get isLoggedIn(): Observable<boolean> {
        return of(true);
      }
    }
    ```

    We use the of() function to return a synchronous Observable that emits a true value for the authentication state.

2. Next, we need to expand our describe() block by defining these variables:

    ```
    let router: Partial<Router>;
    let authService: Partial<AuthService>;
    let routerStateSnapshot: RouterStateSnapshot;
    let activatedRouteSnapshot: ActivatedRouteSnapshot;
    ```

3. Inside the `beforeEach()` function, we must configure the testing
    module by adding the following providers to the `TestBed.`
    `configureTestingModule({})` call:

    ```
    providers: [
      {
        provide: Router,
        useValue: {
          parseUrl: jasmine.createSpy('navigate')
        }
      },
      {
        provide: AuthService,
        useClass: AuthServiceMock
      },
    ]
    ```

4. Next, inject the router and auth services, as follows:

    ```
    router = TestBed.inject(Router);
    authService = TestBed.inject(AuthService);
    ```

    Then, create dummy objects, as follows:

    ```
    routerStateSnapshot = { url: '/feed' } as
    RouterStateSnapshot;
    activatedRouteSnapshot = {} as ActivatedRouteSnapshot;
    ```

    Instead of using the complete objects, we set only the minimal attributes
    required for the test. `CanActivate` accepts `RouterStateSnapshot` and
    `ActivatedRouteSnapshot` objects as arguments, but the auth guard doesn't
    actually use them to allow or deny access. Because of this, we can use these empty
    objects as arguments.

5. Add a second test case, as follows:

    ```
    it('should allow access when user is logged in', () =>
      {});
    ```

    Inside the `it()` block, call the `canActivate()` method of the guard, as follows:

    ```
    const canActivateObs: Observable<boolean | UrlTree> =
      guard.canActivate(
    ```

```
    activatedRouteSnapshot,
    routerStateSnapshot
) as Observable<boolean | UrlTree>;
```

Then, subscribe to the returned Observable and assert that the `isLoggedIn` variable is `true`, as follows:

```
canActivateObs.subscribe((isLoggedIn) => {
    expect(isLoggedIn).toBe(true);
});
```

6. Add the third test case, as follows:

```
it('should redirect to login page when user is logged
out', () => {});
```

Inside the `it()` function, spy on the `isLoggedIn` property of the auth service by executing the following code:

```
spyOnProperty(authService,
    'isLoggedIn').and.returnValue(of(false));
```

Then, call the `canActivate()` method of the guard, as follows:

```
const canActivateObs: Observable<boolean | UrlTree> =
    guard.canActivate(
    activatedRouteSnapshot,
    routerStateSnapshot
) as Observable<boolean | UrlTree>;
```

Finally, subscribe to the returned Observable and assert that the `parseUrl()` method of the router has been called with the `/users/login` path, as follows:

```
canActivateObs.subscribe(() => {
    expect(router.parseUrl)
    .toHaveBeenCalledOnceWith('/users/login');
});
```

Our tests should pass, as shown in the following screenshot:

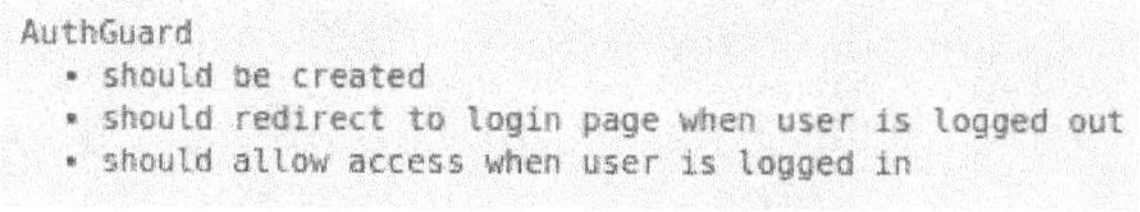

```
AuthGuard
  • should be created
  • should redirect to login page when user is logged out
  • should allow access when user is logged in
```

Figure 8.4 – AuthGuard tests

You can get the code for these test cases at this commit: `https://git.io/J6xBe`.

Testing the header component

Let's start by configuring the tests for the header component before writing the tests. Follow these steps:

1. Open the `core/components/header/header.component.spec.ts` file and include the following imports:

    ```
    import {
      ChangeDetectionStrategy,
      NO_ERRORS_SCHEMA
    } from '@angular/core';
    import { fakeAsync, tick } from
      '@angular/core/testing';
    import { MatDialog } from '@angular/material/dialog';
    import { MatSnackBar } from '@angular/material/
      snack-bar';
    import { By } from '@angular/platform-browser';
    import { Router } from '@angular/router';
    import { defer, of, Subject } from 'rxjs';
    import {
      AuthState,
      SearchUsersResponse,
      UsersResponse } from 'src/app/shared';
    import { AuthService } from 'src/app/core';
    ```

2. Define the following variables in the `describe()` function's callback:

    ```
    let authService: AuthService;
    let authServiceStub: Partial<AuthService>;
    let router: Router;
    let routerStub: Partial<Router>;
    let matSnackBar: MatSnackBar;
    let matDialog: MatDialog;
    ```

3.   You must create a stub for the auth service in the `beforeEach()` function's callback, as follows:

```
authServiceStub = {
  authState: of({
    isLoggedIn: false,
    currentUser: null
  } as AuthState),
  logOut: jasmine.createSpy('logOut'),
  searchUsers: (searchQuery: string, offset: number,
    limit: number): SearchUsersResponse => {
      return {
        data: defer(() => Promise.resolve({
          searchUsers: []
        } as UsersResponse)),
        fetchMore: jasmine.createSpy('fetchMore')
      };
    }
};
```

Then, create a stub for the router service, as follows:

```
routerStub = {
  navigateByUrl: jasmine.createSpy('navigateByUrl')
};
```

4.   The testing module must then be configured by adding the following providers to the object passed to the `TestBed.configureTestingModule()` method:

```
providers: [
  { provide: MatDialog, useValue: {} },
  {
    provide: Router, useValue: routerStub
  },
  {
    provide: MatSnackBar, useValue: {
      open: jasmine.createSpy('open')
    }
  },
```

```
    {
        provide: AuthService, useValue: authServiceStub
    }
]
```

5. Following that, you must configure the testing module to ignore unknown directives by executing the following code:

```
await TestBed.configureTestingModule({
    // [...]
    schemas: [NO_ERRORS_SCHEMA]
})
```

6. Override the header component and set changeDetection to the default strategy rather than the OnPush strategy, which is incompatible with testing. Here's how to do this:

```
await TestBed.configureTestingModule({
// [...]
}).overrideComponent(HeaderComponent, {
    set: {
        changeDetection: ChangeDetectionStrategy.Default
    }
})
    .compileComponents();
```

7. Inject the header component's required services just below the compileComponents() method, as follows:

```
authService = TestBed.inject(AuthService);
router = TestBed.inject(Router);
matSnackBar = TestBed.inject(MatSnackBar);
matDialog = TestBed.inject(MatDialog);
```

After we've configured the tests, let's write test cases for the header component, as follows:

8. Write a first test case, as follows:

```
it('should display the app title', () => {
    const h1Element: HTMLElement = fixture.nativeElement
        .querySelector('h1');
```

```
    expect(hlElement).toBeDefined();
    expect(hlElement.textContent).toContain('ngSocial');
});
```

This asserts that the correct title is rendered.

9. Write a second test case, which asserts that the isLoggedIn property is initially false, as follows:

```
it('should have #isLoggedIn set to false after
construction', () => {
  expect(component.isLoggedIn).toEqual(false);
});
```

10. Write a third test case, which asserts that the authUser property is initially null, as follows:

```
it('should have #authUser set to null after
construction', () => {
  expect(component.authUser).toEqual(null);
});
```

11. Write a fourth test case to assert that the search bar appears if the user is logged in, as follows:

```
it('should display the search bar if user is logged in',
() => {
  component.isLoggedIn = true;
  fixture.detectChanges();
  const searchbar = fixture.debugElement
    .query(By.css('.searchbar'));
  expect(searchbar).toBeTruthy();
});
```

12. Implement the following test case to assert that the component delegated logout to the auth service:

```
it('should logout by delegation to AuthService', () => {
  component.logOut();
  expect(authService.logOut)
    .toHaveBeenCalledTimes(1);
});
```

13. Implement the following test case to assert that the user is directed to the login component after logging out:

```
it('should navigate to /users/login after logout', () =>
{
    component.logOut();
    expect(router.navigateByUrl)
        .toHaveBeenCalledOnceWith('/users/login');
});
```

14. Implement the following test case to assert that the component unsubscribes from any subscriptions after it has been destroyed:

```
it('should unsubscribe when destroyed', () => {
    const nextSpy = spyOn(Subject.prototype, 'next');
    const completeSpy = spyOn(Subject.prototype,
        'complete');
    component.ngOnDestroy();
    expect(nextSpy).toHaveBeenCalledOnceWith(true);
    expect(completeSpy).toHaveBeenCalledTimes(1);
});
```

15. Implement the following test case, which asserts that the component displays a message with a snack bar to notify the user that a search term must be entered before calling the searchUsers() method:

```
it('should open SnackBar if searchInput is empty string',
    () => {
    component.isLoggedIn = true;
    fixture.detectChanges();
    component.searchInput!.nativeElement.value = '';
    component.searchUsers();
    expect(matSnackBar.open)
        .toHaveBeenCalledOnceWith(
            'Please enter a search term',
            'Ok',
            {
                duration: 5 * 1000
            });
});
```

16. Implement the following test case to assert that the component delegated to the auth service's `searchUsers()` method to search for users:

```
it('should delegate to AuthService if searchInput is not
empty string', fakeAsync(() => {
  component.isLoggedIn = true;
  fixture.detectChanges();
  component.searchInput!.nativeElement.value = 'A B';
  spyOn(authService, 'searchUsers')
    .and.callThrough();
  component.searchUsers();
  tick()
  expect(authService.searchUsers)
    .toHaveBeenCalledOnceWith(
      'A B', 0, 10
    );
}));
```

17. Implement the following test case to assert that when the **Search** button is pressed, the component calls the `searchUsers()` method:

```
it('should call searchUsers() when the search button
  is clicked', fakeAsync(() => {
    component.isLoggedIn = true;
    fixture.detectChanges();
      spyOn(component, 'searchUsers');
      const searchBtn = fixture.debugElement
      .query(By.css('button:nth-of-type(1)'));
      searchBtn.nativeElement.click();
        tick();
        expect(component.searchUsers)
          .toHaveBeenCalled();
}));
```

Similarly, as an assignment, implement the following specifications:

- `should call logOut() when the logout button is clicked`
- `should detect changes after ngOnInit call`
- `should emit auth state after subscribe`
- `should display the search button if user is logged in`
- `should display the notifications button if user is logged in`
- `should display the profile button if user is logged in`
- `should display the home button if user is logged in`
- `should display the login button if user is not logged in`
- `should display the logout button if user is logged in`

You can see the complete code for these test cases at `https://git.io/JimqJ`.

In Karma, you should get a similar result, as we can see here:

```
20 specs, 0 failures

HeaderComponent
  • should create
  • should display the app title
  • should have #isLoggedIn set to false after construction
  • should have #authUser set to null after construction
  • should display the search bar if user is logged in
  • should logout by delegation to AuthService
  • should navigate to /users/login after logout
  • should unsubscribe when destroyed
  • should open SnackBar if searchInput is empty string
  • should delegate to AuthService if searchInput is not empty string
  • should call searchUsers() when the search button is clicked
  • should call logOut() when the logout button is clicked
  • should detect changes after ngOnInit call
  • should emit auth state after subscribe
  • should display the search button if user is logged in
  • should display the notifications button if user is logged in
  • should display the profile button if user is logged in
  • should display the home button if user is logged in
  • should display the login button if user is not logged in
  • should display the logout button if user is logged in
```

Figure 8.5 – Tests for the header component

Testing the search dialog component

Following that, you must write some tests for the search dialog component, as follows:

- `should display title with h2 tag`
- `should go to user profile when Visit button is clicked`
- `should close dialog when Close button is clicked`
- `should invoke fetchMore when More.. button is clicked`
- `should emit found users after subscribing to searchUsers`
- `should unsubscribe when component is destroyed`
- `should call navigateByUrl when goToProfile is called`
- `should close dialog on trigger`

You can find the implementation at this commit: `https://git.io/JimHz`.

Testing other components

Also, as assignments, do the following:

1. Write a test case for the footer component to assert that it renders the `ngSocial` `(c) 2021` text with a `<p>` tag.

2. Write a test case for the page-not-found component for asserting that it renders the `This page doesn't exist!` text with a `<p>` tag.

3. Write the following test cases for the signup component:

   `should initially have invalid form`

   `should initially have empty error message`

   `should set error message to 'Please enter valid information'`

   `should set error message to 'Passwords mismatch'`

   `should navigate to user profile when signup succeeds`

   `should display success message when signup succeeds`

   `should display error when signup fails`

   `should unsubscribe when destroyed`

4. Write test cases for the login component, as follows:

    ```
    should initially have invalid form

    should initially have empty error message

    should set error message when email or password is
    not valid

    should navigate to user profile when login succeeds

    should display success message when login succeeds

    should display error when login fails

    should unsubscribe when destroyed
    ```

5. Fix the existing test case for the profile component by adding a stub for the `ActivatedRoute` service.

6. Clean the app component by removing the unused `title` property, and remove the related test cases.

Find the relevant commits at `https://git.io/Jimdh`.

Finally, run the `ng test` command to ensure that all of the tests pass.

Summary

In this chapter, we improved our auth system by adding guards to protect pages that require user auth, then we tested the auth service to ensure it does work as intended.

In the following chapter, we will start implementing the profile component's functionality. We'll add the necessary code to fetch the user that corresponds to a profile URL and render their information on the page, including the ability to upload the user's photo and cover image, as well as adding a biography.

9

Uploading Images and Adding Posts

In the previous chapter, we finished implementing the authentication system by adding the authentication guard and unit tests.

In this chapter, we will start implementing the profile component's functionality. We'll add the necessary code to fetch the user that corresponds to a profile URL and render their information on the page, including the ability to upload the user's photo and cover image, as well as adding a biography.

We'll use extra tools such as GraphQL Code Generator to generate code and automatically infer variable and result types rather than manually specifying them, which keeps our code clean, maintainable, and scalable.

We will cover the following topics:

- Image uploading with Angular and Apollo
- Implementing the post service
- Implementing the profile service
- Implementing the base component
- Implementing the profile component

- Creating the profile UI
- Implementing the `create-post` component

Technical requirements

To complete this chapter's steps, you must first finish the previous chapter of this book.

You should also be acquainted with the following technologies:

- JavaScript/TypeScript
- HTML
- CSS

This chapter's source code can be found at `https://github.com/PacktPublishing/Full-Stack-App-Development-with-Angular-and-GraphQL/tree/main/Chapter07`.

Image uploading with Angular and Apollo

Uploading images is a required feature of any social network; in our case, we need images not only for posts, but also for the user's photo and profile cover.

We've added three mutations to the backend: `uploadFile`, `setUserPhoto`, and `setUserCover`. In this section, we'll write the code for sending requests to call these GraphQL mutations to create posts with or without an image, as well as setting the user's profile photo and cover.

We manually created the GraphQL documents in the previous section by wrapping the queries and mutations with the `gql` tag, which returns the documents (objects of the `DocumentNode` type) that we pass to Apollo Client methods to fetch or mutate data.

Then, for communicating with our GraphQL API, we created types for results/responses and implemented the Apollo services, which extend the generic `Query` or `Mutation` classes.

This works, but as our code base grows, it is unlikely to scale because manually writing and maintaining types requires extra work to ensure the types of the results (and variables) match the selection sets (and GraphQL arguments) we are working with. This increases the likelihood of errors such as typos.

Fortunately, we have generation tools such as GraphQL Code Generator, which generates TypeScript types and even Apollo services.

Using various plugins, this tool generates complete types and code constructs such as GraphQL documents and Apollo services (for apollo-angular) by providing the server schema and GraphQL queries and mutations. Check out `https://www.graphql-code-generator.com`.

To generate TypeScript types for variables and results, we can use the `@graphql-codegen/typescript` and `@graphql-codegen/typescript-operations` plugins.

Then, using the `@graphql-codegen/typescript-apollo-angular` plugin, we can generate Angular services.

This saves you from having to manually create and update the types. The tool was recently expanded so that you don't have to manually specify the types when calling the Apollo Client APIs (or any GraphQL client you prefer to use).

This feature is implemented in the `graphql-typed-document-node` plugin, which generates `DocumentNode` objects with strong typing. See `https://github.com/dotansimha/graphql-typed-document-node` and `https://the-guild.dev/blog/typed-document-node` for more information.

In this section, we'll continue implementing our application, but this time we'll use the code generator, which we'll configure with the necessary plugins, to generate services and typed `DocumentNode` objects to automatically infer variable and result types, rather than specifying them manually, which keeps our code clean, maintainable, and scalable!

Setting up GraphQL Code Generator

Typed document nodes are supported by apollo-angular versions 2.6.0 and higher, so let's make sure we have the most recent versions of the `@apollo/client` and `apollo-angular` packages by running the following commands from the `client/` folder to upgrade the existing dependencies:

```
npm i @apollo/client
npm i apollo-angular
```

Please keep in mind that this step may not be necessary in your case because support for typed documents was added while this book was being written.

Let's go over how to set up GraphQL Code Generator to generate typed documents and services for interacting with the GraphQL API step by step:

1. To begin, open your terminal and execute the following commands from the root of your monorepo project to install the necessary dependencies:

```
npm i @graphql-codegen/cli
npm i @graphql-typed-document-node/core
npm i @graphql-codegen/typescript
npm i @graphql-codegen/typescript-operations
npm i @graphql-codegen/typed-document-node
npm i @graphql-codegen/typescript-apollo-angular
npm i @graphql-codegen/import-types-preset
```

We previously installed the CLI while working on the server-side application, but reinstalling it ensures that we have the most recent version.

2. Open the `codegen.yml` file in the monorepo project's root directory and update it to point to your GraphQL documents using the `documents` field in the root-level configuration, as shown here:

```
documents: './packages/client/src/**/*.graphql'
```

The field was added at the same level as the `schema` field. The field accepts a variety of options. In our case, we used a glob expression to specify multiple files that could exist in the `./packages/client/src/` folder's subfolders. Check out `https://www.graphql-code-generator.com/docs/getting-started/documents-field` for more information on this field.

3. Under the `generates` field, add the following output for generating types:

```
generates:
  # [...]
  ./packages/graphql-types/src/types.ts:
    plugins:
      - typescript
```

To generate the types that will be imported in the next generated files, we use the TypeScript plugin.

4. Similarly, add the following output for generating GraphQL documents and operations:

```
./packages/graphql-types/src/client/
  graphql-operations.ts:
  preset: import-types
  presetConfig:
    typesPath: ../types
  plugins:
    - typescript-operations
    - typed-document-node
```

The `import-types` preset is used to import and reuse previously generated types from the `../types` path, and the `typescript-operations` and `typed-document-node` plugins are used to generate the GraphQL operations and typed documents.

5. Finally, use the `typescript-apollo-angular` plugin to generate the Angular services as follows:

```
./packages/client/src/app/core/gql.services.ts:
  config:
    documentMode: external
    importOperationTypesFrom: Operations
    importDocumentNodeExternallyFrom:
      '@ngsocial/graphql/documents'
  plugins:
    - typescript-apollo-angular
```

In the `config` field, we pass options to import the documents from the `'@ngsocial/graphql/documents'` path, which will be configured to point to the `graphql-operations.ts` file we generated previously, since they are required by the services that will be generated.

Otherwise, we'll have to add the `typescript-operations` plugin here, too, which will generate the same types in the `gql.services.ts` file. Check out more on the `config` field at `https://www.graphql-code-generator.com/docs/getting-started/config-field`.

We use the `importDocumentNodeExternallyFrom` option, with `documentMode` set to `external` to import the base types generated with the `typescript-operations` plugin in their own file, which is `graphql-operations.ts` in our case, and we use the `importOperationTypesFrom` option to prefix them with the `Operations` prefix.

For more information, check out `https://www.graphql-code-generator.com/docs/plugins/typescript-document-nodes`.

This configuration was used to avoid generating the same basic types, document types, and GraphQL operations in each file. We create them once and import them from other files as needed. Check out all the available plugins at `https://www.graphql-code-generator.com/docs/plugins/index`.

Adding fragments and mutations

In this section, we'll add the necessary fragments and mutations for creating posts and uploading images, and then generate the results' types and the Apollo services for sending the mutations to the server, before implementing the `users` module's profile service.

If you are unfamiliar with fragments, read the official guide at `https://graphql.org/learn/queries/#fragments`.

Let's start by adding the GraphQL fragments and mutations:

1. Begin by creating the `graphql.documents/` folder inside the `client/src/app/` folder, then create the `graphql.documents/user.graphql` file, and add the following fragment on the `User` type:

    ```
    fragment BasicUserFields on User {
        id
        fullName
        username
        image
    }
    ```

    This fragment contains a set of basic fields from our GraphQL schema's `User` type.

2. Add the `UserFields` fragment as follows:

    ```
    fragment UserFields on User {
        ...BasicUserFields
        email
        bio
    ```

```
    coverImage
    postsCount
    createdAt
  }
```

This fragment contains the set of all User type fields. To expand the BasicUserFields fragment on the UserFields fragment, we use the . . . (three dots) syntax.

3. Add the setUserPhoto mutation for setting the user's photo:

```
mutation setUserPhoto($file: Upload!) {
  setUserPhoto(file: $file) {
    ...UserFields
  }
}
```

Because this mutation returns the user object whose photo was set, we use a selection set to ask the backend for the exact fields that we need in our frontend app. In the selection set, we expand the UserFields fragment, which refers to the fields of the User type in the schema.

4. Add the setUserCover mutation for setting the user's cover image:

```
mutation setUserCover($file: Upload!) {
  setUserCover(file: $file) {
    ...UserFields
  }
}
```

5. Add the setUserBio mutation for setting the user's bio:

```
mutation setUserBio($bio: String!) {
  setUserBio(bio: $bio) {
    ...UserFields
  }
}
```

Let's now add the necessary Post type fragments.

1. Create the graphql.documents/post.graphql file and begin by adding the following fragment:

    ```
    fragment BasicPostFields on Post {
      id
      text
      image
      createdAt
    }
    ```

2. Add the following fragment, which contains the fields related to the Post type's liking information:

    ```
    fragment LikesInfo on Post {
      likesCount
      latestLike
      likedByAuthUser
    }
    ```

3. Add the following fragment, which contains the fields related to the comments information on the Post type:

    ```
    fragment CommentsInfo on Post {
      latestComment {
        ...CommentFields
        author {
          ...BasicUserFields
        }
      }
      commentsCount
    }
    ```

    We expand the CommentFields fragment that we'll define in the following steps in the latestComment field, which returns an object. We expand the BasicUserFields fragment, which contains the basic fields of the User type, in the author field.

4. Add the `PostFields` fragment as shown:

```
fragment PostFields on Post {
  ...BasicPostFields
author {
    ...BasicUserFields
}
  ...CommentsInfo
  ...LikesInfo
}
```

Next, add three mutations for uploading files, as well as creating and removing posts.

5. Begin by adding the following mutation to the same `post.graphql` file:

```
mutation uploadFile($file: Upload!) {
  uploadFile(file: $file) {
    url
    filename
    mimetype
    encoding
  }
}
```

Add the following mutation for creating posts from text and/or images:

```
mutation createPost($text: String, $image: String) {
  post(text: $text, image: $image) {
    ...BasicPostFields
    author {
      ...BasicUserFields
    postsCount
    }
  }
}
```

We expand the `BasicPostFields` fragment in the selection set, then the `BasicUserFields` fragment in the `author` field.

Add the following mutation to remove a post by ID:

```
mutation removePost($id: ID!) {
  removePost(id: $id)
}
```

6. Create the `graphql.documents/comment.graphql` file as well, and add the following fragment to the GraphQL schema's `Comment` type:

```
fragment CommentFields on Comment {
  id
  comment
  createdAt
}
```

We'll see how to generate the types and Apollo services using the code generator now that we've added the fragments and mutations required for uploading images, creating, and removing posts.

Generating types and services

Let's use the code generator to generate the types and services:

1. To start the code generator from the root of the monorepo project, use the following command:

```
npm run codegen
```

This will generate the relevant files, which will contain the appropriate code and types. In the `@ngsocial/graphql` package, the types and documents are generated, while the Angular services are generated in the `core/` folder of the client project.

2. Open the `client/tsconfig.json` file and add the following paths to the generated files:

```
"paths": {
  "@ngsocial/graphql": ["../graphql-types/src"],
  "@ngsocial/graphql/types": ["../graphql-
    types/src/types"],
  "@ngsocial/graphql/documents": ["../graphql-
    types/src/client/graphql-operations"]
}
```

Instead of specifying the full paths, we will be able to import the types and document node objects from the `@ngsocial/graphql/types` and `@ngsocial/graphql/documents` paths.

Implementing the post service

Let's now generate and implement the post service, which encapsulates the methods for creating and removing posts, as well as uploading images, after we've generated the types and services and configured the paths. Simply inject the generated `CreatePostGQL`, `RemovePostGQL`, and `UploadFileGQL` services and call their `mutate()` method with the necessary arguments to send the mutations to the backend:

1. To generate a post service in the `core/` folder, open your terminal and run the following command:

    ```
    ng g s core/services/post/post
    ```

    Next, create the `core/services/index.ts` file and export the existing services to the import paths from within the `core/` folder:

    ```
    export { AuthService } from './auth/auth.service';
    export { PostService } from './post/post.service';
    ```

    Also, open the `core/index.ts` file and update it as follows:

    ```
    export {
        AuthService,
        PostService
    } from './services';
    export { AuthGuard } from './guards/auth/auth.guard';
    export * from './gql.services';
    ```

    We re-export the post service and all of the Apollo services, which were generated by the code generator in the `./gql.services.ts` file.

2. Add the following imports to the `core/services/post/post.service.ts` file:

    ```
    import { map } from 'rxjs/operators';
    import {
      CreatePostGQL,
      RemovePostGQL,
      UploadFileGQL
    } from 'src/app/core';
    ```

3. Next, use the service constructor to inject the imported services as follows:

```
export class PostService {
  constructor(
    private createPostGQL: CreatePostGQL,
    private removePostGQL: RemovePostGQL,
    private uploadFileGQL: UploadFileGQL) { }
```

4. Define and implement the following file uploading method:

```
uploadFile(image: File) {
  return this.uploadFileGQL.mutate(
    {
      file: image
    },
    {
      context: {
        useMultipart: true
      }
    }).pipe(map(result => result.data!.uploadFile));
}
```

5. Define and implement the following method for creating posts:

```
createPost(text: string | null, image: string | null) {
  return this.createPostGQL
    .mutate({
      text: text,
      image: image
    })
    .pipe(map(result => result.data!.post));
}
```

6. Define and implement the method for removing posts by ID as follows:

```
removePost(id: string) {
  return this.removePostGQL
    .mutate({
      id: id
    })
```

```
    .pipe(map(result => result.data!.removePost));
}
```

We've added the necessary methods for creating and removing posts, as well as uploading post images. Following that, we must implement the profile service, which contains the code for configuring the user's bio, photo, and cover. Then, we'll look at how to use the profile component to add a UI for calling these.

Implementing the profile service

In this section, we'll build the profile service, which contains the code for interacting with the user's profile. This service injects the generated `SetUserBioGQL`, `SetUserPhotoGQL`, and `SetUserCoverGQL` services and implements the `setUserBio()`, `setUserPhoto()`, and `setUserCover()` methods, which simply send the necessary mutations to the backend:

1. Import the following symbols into the `users/services/profile/profile.service.ts` file (which was created in the previous chapter):

    ```
    import {
      SetUserCoverGQL,
      SetUserPhotoGQL,
      SetUserBioGQL
    } from 'src/app/core';
    import { map } from 'rxjs/operators';
    ```

    Using the constructor, inject Apollo services:

    ```
    export class ProfileService {

      constructor(
        private setUserBioGQL: SetUserBioGQL,
        private setUserPhotoGQL: SetUserPhotoGQL,
        private setUserCoverGQL: SetUserCoverGQL
      ) { }
    ```

2. Implement the `setUserBio()` method as follows:

    ```
    setUserBio(bio: string) {
      return this.setUserBioGQL.mutate({
        bio: bio
    ```

```
}).pipe(map(result => result.data!.setUserBio));
}
```

3. Implement the `setUserPhoto()` method as follows:

```
setUserPhoto(photoImage: File) {
  return this.setUserPhotoGQL.mutate({
    file: photoImage
  },
  {
    context: {
      useMultipart: true
    }
  }).pipe(map(result => result.data!.setUserPhoto));
}
```

4. Implement the `setUserCover()` method as follows:

```
setUserCover(coverImage: File) {
  return this.setUserCoverGQL.mutate({
    file: coverImage
  },
  {
    context: {
      useMultipart: true
    }
  }).pipe(map(result => result.data!.setUserCover));
}
```

Implementing the base component

Because the implementations of the user's profile and the feed's posts components are similar, a substantial amount of code is shared between these components. To avoid reinventing the wheel and adhere to the **Don't Repeat Yourself** (**DRY**) principle, we can use TypeScript inheritance with an abstract base class to implement shared functionality and then extend it from both components.

Let's go over how to implement the profile component step by step:

1. Create the `shared/types/post.event.ts` file and add the following type:

    ```
    export type PostEvent = {
      text: string | null;
      image: File | null;
    };
    ```

    Then, create the `shared/types/removepost.event.ts` file and add the following type:

    ```
    export type RemovePostEvent = {
      id: string;
    };
    ```

    These types are intended for the typing of the custom events that will be used to communicate between the parent, profile, and posts components (that extend the base class), and the child components that are in charge of creating and removing posts. These components will be added later, but the methods that handle the events will be added to the base class in the next steps.

    Open the `shared/index.ts` file and export the types as follows:

    ```
    export { PostEvent } from './types/post.event';
    export { RemovePostEvent } from
        './types/removepost.event';
    ```

2. Create the `core/components/base.component.ts` file and add the following imports:

    Start by importing the `Injector` and life cycle interfaces:

    ```
    import {
      Injector,
      Component,
      OnDestroy,
      OnInit
    } from '@angular/core';
    ```

Import the following `rxjs` symbols and the `snack-bar` component:

```
import {
  Subject
} from 'rxjs';
import {
  mergeMap,
  takeUntil
} from 'rxjs/operators';
import { MatSnackBar }
  from '@angular/material/snack-bar';
```

Import the authentication and post services:

```
import {
  AuthService,
  PostService
} from '../services';
```

Import the `User` and `Post` types:

```
import {
  User,
  Post
} from '@ngsocial/graphql/types'
```

In this case, we're importing the `User` type generated by the code generator rather than the `User` model we created manually.

Finally, import the custom event types:

```
import {
  PostEvent,
  RemovePostEvent
} from 'src/app/shared';
```

3. Define the following abstract class, including its public, protected, and private attributes:

```
@Component({
  template: ''
})
export abstract class BaseComponent
  implements OnInit, OnDestroy {}
```

Define the following members in the class's body:

```
public authService: AuthService;
public postService: PostService;
public posts: Post[] = [];
public authUser: Partial<User> | null = null;
public loading: boolean = false;
protected snackBar: MatSnackBar;
private componentDestroyed = new Subject();
```

We defined the properties that will be used in the parent components, such as the `posts` array, which will hold the fetched posts, the `authUser` object, which will hold the authenticated user, and the `loading` property, which will track the loading state of operations.

We also defined the `componentDestroyed` subject, which will be used to unsubscribe from the returned observables of various methods after the component has been destroyed.

Finally, we defined the `authService`, `postService`, and `snackBar` properties, which will be used to assign dependencies with the `Injector` service rather than injecting them via the constructor to simplify the child components.

The private attributes are only accessible from within the base component, whereas the protected attributes are accessible from both the base and child components. The public properties are accessible from within the components and the associated templates.

4. Next, instead of using the constructor's arguments, let's inject the dependencies inside the constructor:

```
constructor(public injector: Injector) {
    this.snackBar = injector.get(MatSnackBar);
    this.authService = injector.get(AuthService);
    this.postService = injector.get(PostService);
}
```

We inject the `Injector` service in the constructor and then use it to get the required dependencies. This eliminates the need for the child components to re-pass all of the parent dependencies into their constructor. For more information, check out https://devblogs.microsoft.com/premier-developer/angular-how-to-simplify-components-with-typescript-inheritance/.

5. Use the following code to implement the life cycle's ngOnOnit() method:

```
ngOnInit(): void {
  this.authService
    .authUser
    .pipe(
      takeUntil(this.componentDestroyed))
    .subscribe({
      next: (authUser) => {
        this.authUser = authUser;
      }
    });
}
```

We also assign the authenticated user to the component's authUser property so that we can access it without referencing the authentication service in our component and its children.

6. Use the following code to implement the ngOnDestroy() life cycle method:

```
ngOnDestroy(): void {
  this.componentDestroyed.next();
  this.componentDestroyed.complete();
}
```

If the component is destroyed, we emit a value from the componentDestroyed subject and complete the subject. This will serve as a notifier for the takeUntil() operator, which will be applied to the methods we'll implement in the following steps to unsubscribe from the returned observables. When our subject emits a value, the takeUntil() operator unsubscribes from the source observable to which we are subscribed.

7. To handle errors, add the protected handleErrors() method, which will be passed to the subscribe() method's error handler:

```
protected handleErrors(err: Error) {
  if(this.loading) this.loading = false;
  console.log(err);
  this.snackBar.open(err.message, 'Ok', {
    duration: 5 * 1000
```

```
        });
    }
```

When an error occurs, the `loading` property is set to `false`, and the error message is displayed. The `loading` property controls whether or not a loading indicator appears in the user interface.

8. Add the `displayMessage()` method, which uses a snack bar to display the message passed as an argument:

```
protected displayMessage(message: string) {
    this.snackBar
        .open(message,
            'Ok', {
            duration: 5 * 1000
        });
}
```

9. Add the `createPost()` method to either create a text-only post or display an error message, as shown here:

```
private createPost(text: string | null) {
    if (!text?.length) {
        this.displayMessage('Cannot create an empty
            post.');
        return;
    }
    if (text) {
        this.createTextPost(text);
    }
}
```

To create a post without an image, this method will be called. If the text passed as an argument is empty, a **Cannot create an empty post.** error is displayed with the snack bar component, and the method is exited. Otherwise, we delegate the action to the next step's `createTextPost()` method.

10. Implement the `createTextPost()` method, which passes the creation of the post to the post service:

```
private createTextPost(text: string): void {
    this.loading = true;
    this.postService
        .createPost(text, null)
        .pipe(
          takeUntil(this.componentDestroyed))
        .subscribe(/* ... */);
}
```

When the operation is about to begin, we set the `loading` property to `true`, and when the operation succeeds or fails, we set its value to `false`. To avoid memory leaks, we'll also use the `takeUntil()` operator to unsubscribe from the returned observable.

Then, pass the following observer object to the `subscribe()` method:

```
{
    next: (post) => {
      if (post) {
          this.loading = false;
          this.displayMessage('Your post was created.');
          console.log(post);
      }
    },
    error: (err) => this.handleErrors(err)
}
```

If the operation is successful, we set the `loading` property to `false` and use the snack bar component to display a success message. Otherwise, we handle the error.

11. Implement the `uploadImageAndCreatePost()` method to upload an image and create a post with or without text:

```
private uploadImageAndCreatePost(image: File, text:
    string | null) {
    this.loading = true;
    this.postService
        .uploadFile(image)
```

```
    .pipe(
      mergeMap(uploadFile => this.postService
        .createPost(text, uploadFile?.url || null)
      ),
      takeUntil(this.componentDestroyed)
    ).subscribe(/* ... */);
}
```

In this method, we begin by setting the `loading` property to `true`, and then call the post service's `uploadFile()` method and subscribe to the returned observable.

We use the `mergeMap()` operator to chain the observable for creating the post with the observable to upload the image because the call to the `createPost()` method is dependent on the result of the `uploadFile()` method. When we subscribe to the combined observable, it will combine the two observables into one and ensure that the inner observable emits after the source.

When the source observable emits the uploaded file's information, the emitted value is passed on to the function, which we passed to the `mergeMap()` operator, where we call the `createPost()` method, which returns the inner observable.

This means that the image will be uploaded first, and the output will be passed to the method in charge of creating the post. The order is critical because we must first obtain the URL of the uploaded image before calling the `createPost()` method.

The `mergeMap()` operator maps and flattens each value emitted from the source observable, which is returned by the `uploadFile()` method, to the inner observable, which is returned by the `createPost()` method.

See `https://rxjs.dev/api/operators/mergeMap`.

We also use the `pipe()` method to apply the `takeUntil()` operator to the source observable before calling the `subscribe()` method.

We pass the `componentDestroyed` subject as a notifier to the `takeUntil()` operator to unsubscribe from the observable when the component is destroyed.

Then, pass the following observer object to the `subscribe()` method:

```
{
  next: (post) => {
    this.loading = false;
    if (post) {
```

```
        this.displayMessage('Your post was created.');
        console.log(post);
      }
    },
    error: (err) => this.handleErrors(err)
  }
```

If the post is successfully created, we display a success message, otherwise, we handle the error. To handle errors in our method, we call the `handleErrors()` method from within the error handler.

12. Implement the following `onPost()` method, which is called when the user wants to create a post:

```
onPost(e: PostEvent): void {
  if (e.image) {
    this.uploadImageAndCreatePost(e.image, e.text);
  }
  else {
    this.createPost(e.text);
  }
}
```

This method will be invoked in response to a custom event for post creation. If the user has chosen an image, we use the `uploadImageAndCreatePost()` method to upload the image and create the post, which can be with or without text (the `e.text` object can be `null`). Otherwise, we'll use the `createPost()` method to create a text-only post.

13. Add the `onRemovePost()` method to remove a post based on its ID:

```
onRemovePost(e: RemovePostEvent): void {
  this.postService
    .removePost(e.id)
    .pipe(takeUntil(this.componentDestroyed))
    .subscribe({
      next: () => {
        this.displayMessage('Post deleted.');
      },
      error: (err) => this.handleErrors(err)
```

```
    });
  }
```

This method will be called in response to a custom event for removing a specified post by ID. It calls the `removePost()` method of the injected post service and subscribes to the returned observable. If the operation is successful, we display a success message that says `Post deleted.`. Otherwise, we handle the error.

We use the `takeUntil()` operator with the `componentDestroyed` subject to unsubscribe from the returned observable when the component gets destroyed.

14. Finally, let's add the base component to the `core/` folder's barrel file. Add the following line to the `core/index.ts` file:

```
export { BaseComponent }
    from './components/base.component';
```

As we progress through the book, we'll add more methods to the base component, but for now, this is all that's needed to move on to the next steps of the profile component's implementation.

Implementing the profile component

In this section, we'll implement the profile component by extending the base component:

1. To begin, open the `core/index.ts` file and export the profile service using the following syntax:

```
export { ProfileService }
    from '../users/services/profile/profile.service';
```

2. Open the `users/components/profile/profile.component.ts` file and add the imports as follows:

```
import {
  Component,
  ElementRef,
  Injector,
  ViewChild
} from '@angular/core';
import {
  ActivatedRoute,
  ParamMap
```

```
} from '@angular/router';
import { User }
  from '@ngsocial/graphql/types';
import {
  User
    as UserModel
} from 'src/app/shared';
import { Maybe } from 'graphql/jsutils/Maybe';
import { Subject } from 'rxjs';
import {
  take,
  takeUntil,
  switchMap
} from 'rxjs/operators';
import {
  BaseComponent
} from 'src/app/core/components/base.component';
import { authState }
  from 'src/app/reactive';
import { ProfileService }
  from 'src/app/core';
```

3. Update the component's class as follows:

```
export class ProfileComponent
  extends BaseComponent {
  private destroyNotifier: Subject<boolean> = new
    Subject();
  @ViewChild('bioInput') bioInput: ElementRef | null =
    null;
  profileUser: Partial<User> | null = null;
  showEditSection: boolean = false;
  isAuthUserProfile: boolean = false;
  constructor(
    private route: ActivatedRoute,
    private profileService: ProfileService,
    public injector: Injector
```

```
) {
    super(injector);
}
}
```

We define the public properties that will be accessed from the template, as well as the private members that will only be used in the component's class:

- The `profileUser` property, which contains the user of the profile

- The `showEditSection` Boolean, which specifies whether the profile's editing section is visible or hidden

- The `isAuthUserProfile` Boolean, which indicates whether the profile belongs to the authenticated user

- The `bioInput` element reference, decorated with `ViewChild`, which will be used to get the value of the bio input field in the template

- The `destroyNotifier` subject, which notifies the `takeUntil` operator when the component is destroyed

We inject the required services, such as `Injector`, `ActivatedRoute`, and `ProfileService`, into the constructor. The injector is passed to the base service, and the profile component makes use of `ActivatedRoute` and `ProfileService`.

4. Define the following setter and getter properties:

The following code block is used to set the value of the user's bio textarea:

```
set bioInputValue(value: string) {
    this
        .bioInput!
        .nativeElement
        .value = value;
}
```

The following code block is used to get the value of the bio textarea:

```
get bioInputValue() {
    return this
        .bioInput!
        .nativeElement
        .value;
}
```

The following code block is used to obtain the profile user's first name:

```
get userFirstName() {
  return this.profileUser
    ?.fullName
    ?.split(' ')
    ?.shift();
}
```

We first use the `split()` method to split the name by space, which returns an array of two strings, and then we use the array's `shift()` method to get the first element, which is the profile user's first name.

5. Make the following changes to the `ngOnInit()` method:

```
ngOnInit(): void {
  super.ngOnInit();
  const userObs = this.route.paramMap
    .pipe(
      switchMap((params: ParamMap) => {
        const userId = params.get('userId')!;
        return this.authService.getUser(userId);
      }),
      takeUntil(this.destroyNotifier)
    );
  userObs.subscribe({
    next: (userResponse) => {
      this.setProfileUser(userResponse.getUser);
      this.setIsAuthUserProfile();
    },
    error: (err) => super.handleErrors(err)
  });
}
```

We use the `ActivatedRoute` service to get the user ID from the profile URL, then we call the authentication service's `getUser()` method (manually injected into the base component), and subscribe to the returned observable to get the user identified by the user ID obtained.

If the operation is successful, we invoke the private `setProfileUser()` and `setIsAuthUserProfile()` methods, which simply assign the profile user from the response body to the component's `profileUser` property and set a Boolean attribute indicating whether the profile belongs to the authenticated user.

In the event of an error, we call the base component's `handleErrors()` method. To avoid memory leaks in our application when the component is destroyed, we use the RxJS `takeUntil()` operator to automatically unsubscribe from the returned observable.

6. Define the private `setUserProfile()` and `setIsAuthUserProfile()` methods:

```
private setProfileUser(user: Partial<User>) {
  this.profileUser = user;
}
private setIsAuthUserProfile() {
  this.isAuthUserProfile =
    !!(
      this.authUser &&
      this.authUser.id == this.profileUser?.id
    );
}
```

7. Define the `setBio()` method, to set the user's biography, as follows:

```
setBio(): void {
  const bio = this.bioInputValue;
  this.profileService
    .setUserBio(bio)
    .pipe(take(1))
    .subscribe(/* ... */);
}
```

Then, pass the following observer object to the `subscribe()` method:

```
{
  next: (userResult) => {
    if (userResult) {
      this.handleSuccess(
        userResult,
```

```
         'Your bio is updated.'
      );
      this.bioInputValue = '';
   }
  },
  error: (err) => this.handleErrors(err)
}
```

8. Define the `setUserCover()` method, to set the user's profile cover, as follows:

```
setUserCover(file: File): void {
  this.profileService
    .setUserCover(file)
    .pipe(take(1))
    .subscribe(/* ... */);
}
```

Then, pass the following observer object to the `subscribe()` method:

```
{
  next: (userResult) => {
    if (userResult)
      this.handleSuccess(
        userResult,
        'Your profile cover is successfully updated.'
      )
  },
  error: (err) => this.handleErrors(err)
}
```

9. Define the `setUserPhoto()` method, to set the user's photo, as follows:

```
setUserPhoto(file: File): void {
  this.profileService
    .setUserPhoto(file)
    .pipe(take(1))
    .subscribe(/* ... */);
}
```

Then, pass the following observer object to the subscribe() method:

```
{
  next: (userResult) => {
    this.handleSuccess(
      userResult,
      'Your profile photo is successfully updated.'
    )
  },
  error: (err) => this.handleErrors(err)
}
```

10. Define the private handleSuccess() method:

```
private handleSuccess(
  userResult: Partial<Maybe<User>>,
  message: string) {
  if (userResult) {
    this.displayMessage(message);
    const prevAuthState = authState();
    authState({
      ...prevAuthState,
      currentUser: userResult as UserModel
    });
    if (this.authUser) {
      this.authUser = userResult as User;
      this.authService
        .storeUser(this.authUser as any);
    }
  }
}
```

11. Define the following method for displaying/hiding the profile's editing section by setting the showEditSection property to true/false:

```
enableDisableEditing(): void {
  this.showEditSection = !this.showEditSection;
}
```

Define the following method, which will be invoked when a user selects a photo from the system file dialog:

```
onPhotoSelected(event: Event): void {
  const files: FileList =
    (event.target as HTMLInputElement)
      .files!;
  if (files.length > 0) {
    const photoFile = files[0];
    this.setUserPhoto(photoFile);
  }
}
```

We get the image file from the event object and use our component's `setUserPhoto()` method to upload it.

12. Finally, define the method that will be called when the user chooses a photo for their profile cover:

```
onCoverSelected(event: Event) {
  const files: FileList =
    (event.target as HTMLInputElement)
      .files!;
  if (files.length > 0) {
    const coverFile = files[0];
    this.setUserCover(coverFile);
  }
}
```

This method, like the previous one, is called when the user selects a cover image from their system file dialog. We get the cover image from the event object and use `setUserCover()` to upload it to the server.

Before moving on to the next section, we need to change one behavior in our authentication service: we need to remove the current user and token from `localStorage` when the token expires.

Simply open the `core/services/auth/auth.service.ts` file and make the following changes to the constructor:

```
constructor(
  /* ... */
```

```
) {
  const localToken = this.getLocalToken();
  let isLoggedIn = false;
  if (localToken) {
    isLoggedIn = this.tokenExists() && !this.
tokenExpired(localToken);
  }
  if(!isLoggedIn){
    localStorage.removeItem(ACCESS_TOKEN);
    localStorage.removeItem(AUTH_USER);
  }
  // [...]
}
```

If the token expires, the `isLoggedIn` variable contains a false value. In this case, the access token and authentication user are removed from `localStorage`.

Defining field policies

Before we begin building the UI, we must first apply some transformations to the user's fields before they are displayed on the profile UI. We can accomplish this in a variety of ways, including using built-in or custom Angular directives and pipes, as well as Apollo Client's type and field policies.

Let's look at how we can use field policies to transform the user's `coverImage` and `createdAt` fields:

1. Open the `client/src/app/cache.ts` file and add the `User` type policy to the in-memory cache's `typePolicies` object as follows:

    ```
    User: {
        fields: {}
    }
    ```

2. Then, in the `fields` object, add the fields to be transformed, beginning with the `coverImage` field policy:

    ```
    coverImage: {
        read(coverImage) {
            return 'url(${coverImage})';
    ```

```
        }
    }
```

When accessing the field, this will return the coverImage URL wrapped in the CSS url() function, which can be assigned to the background-image style attribute of the HTML element where we will display the cover image.

3. Then, add the createdAt field policy:

```
createdAt: {
    read(createdAt) {
        return new Date(Number(createdAt))
            .toLocaleDateString('en-US', {
                weekday: 'short',
                day: 'numeric',
                year: 'numeric',
                month: 'long'
            });
    }
}
```

When accessing the createdAt field, this will return the corresponding human-readable date.

Creating the profile UI

Let's build the profile UI after we've implemented the component's methods:

1. Start by adding the following HTML markup to the users/components/profile/profile.component.html file:

```
<mat-card class="profile-cover-card">
<mat-card-content *ngIf="profileUser">
  <div class="profile-cover" [ngStyle]="
    {'background-image': profileUser.coverImage }">
  </div>
  <img class="profile-image" src="{{ profileUser.image
    }}">
</mat-card-content>
</mat-card>
```

To display the profile cover and image, we use a Material Card component. We use the `ngStyle` directive to set the CSS `background-image` property to the URL of the `profileUser` cover photo, and we show the user's photo via interpolation using the element's `src` attribute.

2. Just below the previous card, add the following markup:

```
<mat-card class="profile-call-action">
<mat-card-content *ngIf="!authUser">
  <h1>Do you know {{ userFirstName }}?</h1>
  <p>
   If you know {{ userFirstName }}, join our network
     to see their posts.
  </p>
</mat-card-content>
</mat-card>
```

This is another card with a call to action for the visiting user to join our network. We conditionally render this card with the `*ngIf` directive so that it appears only if the visitor is not registered or logged in and if we have a profile with the user ID retrieved from the visited URL.

3. To get the first name from the user's full name, we use the `userFirstName` getter. Add the following markup directly below the previous card:

```
<div *ngIf="profileUser; else userNotFound"
    class="profile-grid">
<mat-card>
  <mat-card-title>
   <h1> {{ profileUser.fullName }} </h1>
  </mat-card-title>
  <mat-card-content>
   <!-- ... -->
  </mat-card-content>
</mat-card>
<div class="posts-column">
<!-- ... -->
</div>
</div>
```

4. Inside the `<mat-card-content>` element, add the following markup:

```
<p>
  <mat-icon>person</mat-icon>
  <span>
   {{ profileUser.bio }}
  </span>
</p>
<p>
  <mat-icon>calendar_today</mat-icon>
  <span>
   Member since {{ profileUser.createdAt }}
  </span>
</p>
<p>
  <mat-icon>article</mat-icon>
  <span>
   {{ profileUser.postsCount }} posts.
  </span>
</p>
<button *ngIf="isAuthUserProfile" mat-raised-button
  color="primary" class="profile-bio-btn"
  (click)="enableDisableEditing()">
  Edit..
</button>
<!-- ADD MARKUP FOR PROFILE UPDATE SECTION HERE -->
```

5. Just below the previous markup, add the following markup for the profile update section:

```
<div class="profile-settings" *ngIf="showEditSection
  && isAuthUserProfile">
  <button mat-raised-button color="primary"
    class="profile-bio-btn"
      (click)="uploadCover.click()">
    Upload cover
  </button>
  <input #uploadCover type='file' style="display:none"
```

```
        (change)="onCoverSelected($event)" />
    <button mat-raised-button color="primary"
      class="profile-bio-btn"
        (click)="uploadPhoto.click()">
      Upload photo
    </button>
    <input #uploadPhoto type='file' style="display:none"
        (change)="onPhotoSelected($event)" />
    <div class="bio-box-wrapper">
      <textarea #bioInput class="profile-bio-box"
        placeholder="Write a bio.."></textarea>
    </div>
    <button mat-raised-button color="primary"
        class="profile-bio-btn" (click)="setBio()">
    Add a new bio..</button>
  </div>
```

Add the following template, which will be displayed if no users are found:

```
<ng-template #userNotFound>
  <div class="user-not-found">
    <p>
      User not found!
    </p>
  </div>
</ng-template>
```

The CSS styles for styling the profile component will be added in the following section.

Styling the profile component

Add the following styles to the `users/components/profile/profile.component.css` file to style the profile component:

1. Begin by including the following styles:

    ```
    * {
      box-sizing: border-box;
    }
    mat-card {
    ```

```
        margin: 0.4rem;
    }
    .profile-cover-card {
        margin: 0px;
        padding: 0px;
    }
```

These styles are used to set box-sizing to border-box, which instructs the browser to take any border and padding into account when calculating the width and height values we give to our elements.

Next, we set the margin for the Material Card elements, and we set the margin and padding for the card containing the profile cover image and user's photo to 0 pixels each.

2. Style the element(s) with the .profile-cover class (the <div> element containing the cover image) as follows:

```
    .profile-cover {
        height: 10rem;
        width: 100%;
        background-position: center;
        background-repeat: no-repeat;
        background-size: cover;
        background-color:rgb(97, 76, 76);
        position: absolute;
    }
```

To take up the entire width of the page, we set the width to 100%, and then we set the height to 10rem. The background's properties are then set to center for the position and no-repeat for the repeat attribute. The most important thing is that we set the background-size property to cover and use the absolute positioning.

Check out https://developer.mozilla.org/en-US/docs/Web/CSS/ CSS_Backgrounds_and_Borders/Resizing_background_images.

3. Add the following styles of the user's image:

```
.profile-image {
  width: 9rem;
  height: 9rem;
  clip-path: circle(60px at center);
  background-color: gray;
  border-width: 1px;
  margin-top: 5rem;
}
```

In this case, we set the width and height to the same value of 9rem, clip the image to a circle using the clip-path property, and set a default background for the user avatar that will be displayed if the user does not upload a photo.

Check out https://developer.mozilla.org/en-US/docs/Web/CSS/clip-path.

4. Next, let's style the element with CSS Grid layout:

```
.profile-grid {
  display: grid;
  grid-template-columns: 1fr 2.5fr 1.5fr;
}
.profile-bio-btn {
  margin: 0.4rem;
  width: 100%;
}
.profile-bio-box {
  width: 100%;
  height: 5rem;
}
.bio-box-wrapper > textarea {
  padding: 10px;
  background-color: #f0f2f5;
  border-radius: 9px;
}
```

The `profile-grid` class includes all of the information below the cover image and user's photo, as well as the elements containing the full name, bio, date, post counts, and posts.

We switch the display to grid mode and divide the available horizontal space into three columns with ratios of 1, 2.5, and 1.5 using the `fr` unit (the fractional unit). Check out `https://www.techiediaries.com/css-grid-layout-tutorial/` for a detailed explanation.

5. Style the element that will be displayed if no user matching the corresponding profile user ID is found:

```
.user-not-found {
    text-align: center;
    font-size: 5vw;
    margin: 100px;
}
```

6. For small screens with a maximum width of `600px`, add the following styles:

```
@media (max-width: 600px) {
    .profile-grid {
        display: grid;
        grid-template-columns: 1fr;
    }
}
```

If the screen width is less than `600px`, we only use one column for our profile grid element, which takes up all of the available horizontal space.

After you've added these styles, you'll have a profile that looks as in the following screenshot after creating an account and adding both the cover and photo images:

Figure 9.1 – The top section of the profile component

The following is a screenshot of the user's bio section after clicking the **Edit..** button:

Figure 9.2 – User's bio section

The following is a screenshot of the user's bio section after adding a bio:

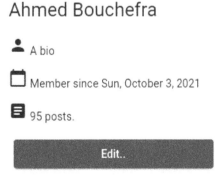

Figure 9.3 – User's information section

Now that we've finished implementing the user's information section of our profile component, let's move on to the user's posts section.

Implementing the create post component

In this section, we'll implement the `create post` component, which will be used to create posts from the profile page (and also from the home feed). We'll put the logic for creating posts in its own reusable component because it will be required on multiple pages.

This component is known as a **presentational** or **dumb** component. It receives input and emits custom output events to the parent component. This type of component has no idea how to fetch data (no injected services), so its only responsibility is to receive data, render it, and emit some events for the parent component(s) to handle. These parent components in our case are the profile and feed's posts components, which are referred to as the **container** or **smart** components.

When this component's `post` event fires, the `onPost()` method of the base component that we defined earlier will be called. This custom event will be triggered when the user clicks on a button for creating posts, which we will implement in the following steps:

1. Begin by generating a component by running the following command from your `client/` folder:

    ```
    ng g c core/components/create-post --module=shared
    ```

2. Export the `create-post` component from the `shared/shared.module.ts` file as follows:

    ```
    @NgModule({
        declarations: [CreatePostComponent],
        imports: [
          CommonModule, ReactiveFormsModule, ...matModules
        ],
        exports: [ReactiveFormsModule, ...matModules,
          CreatePostComponent]
    })
    ```

3. Start by adding the following imports to the `create-post.component.ts` file in the `core/components/create-post/` folder:

    ```
    import {
      Component,
      ElementRef,
      EventEmitter,
      Input,
    ```

```
    OnInit,
    Output,
    ViewChild
} from '@angular/core';
import { User }
    from '@ngsocial/graphql/types';
import { PostEvent }
    from 'src/app/shared';
```

We import the Angular core module's APIs, the User type, and PostEvent we defined previously.

4. Define the component's public and private properties as follows:

```
@Input() authUser!: Partial<User>;
@Input() loading: boolean = false;
@Output() post: EventEmitter<PostEvent> = new
    EventEmitter();
public imageFile: File | null = null;
@ViewChild('postText') postText!: ElementRef;
```

We add two custom input properties for receiving the authenticated user and loading state, via the parent component, as well as a custom event that will be fired when users click the button to create a post.

Using input properties and events, we ensure parent-child communication between components. For more information, see https://angular.io/guide/inputs-outputs#sending-data-to-a-child-component.

We decorate the class properties with the @Input decorator to create input properties, and we can create a custom event by creating an instance of EventEmitter and exposing it as a property decorated with @Output that parent components can bind to, to listen for the custom event and access the associated data via a $event object.

TypeScript will throw errors if we do not initialize all of the defined component properties with default values. This is because Angular 12+, by default, enables the strict mode's strictPropertyInitialization rule.

We also define the `imageFile` property, which contains the user's selected image if it exists, and the `postText` property, which we decorated with `ViewChild` to allow us to get the text value of the input element (referenced by the `#postText` template reference) in the associated template:

1. Define the following getter to obtain the first name of the authenticated user:

```
get userFirstName () {
  return this.authUser
    ?.fullName
    ?.split(' ')
    ?.shift();
}
```

2. Define the following method to obtain the user's selected image:

```
onFileSelected (event: Event): void {
  const files: FileList =
    (event.target as HTMLInputElement)
      .files!;
  if (files.length > 0) {
    this.imageFile = files[0];
  }
}
```

This method will be bound to the `change` event of the input field used to select an image from the user's computer. The image file information is extracted from the event and stored in the `imageFile` property.

3. Define the following method for handling the button's click event to create a post:

```
handlePostClick (): void {
  if (this.authUser!.id) {
    this.post.emit({
      text: this.postText?.nativeElement.value,
      image: this.imageFile as File
    });
  }
  if (this.postText.nativeElement) {
    this.postText.nativeElement.value = '';
  }
}
```

```
      this.imageFile = null;
    }
```

Because this child component is only for presentation and does not communicate with the service, we must emit a custom event to the parent component via the `emit()` method of `EventEmitter`, which is available on the component's `post` property, and the parent component will take care of actually creating the post via the `onPost()` method by delegating the operation to the injected post service.

The `emit()` method of `EventEmitter` sends data to the parent component, in this case, the `PostEvent` object, which contains the text and image required for the parent component and the delegated service to create a post.

After the event is fired, we use the `nativeElement.value` properties of `ElementRef` to clear the value of the `<textarea>` element referenced by the `postText` property, and we set the `imageFile` property to `null`.

4. Following that, we'll add the markup for the component's template. Open the `create-post.component.html` file and add the following changes:

```
<mat-card class="post-box-wrapper">
<mat-card-content>
  <mat-progress-bar style="margin-bottom: 20px;"
    mode="indeterminate" *ngIf="loading"></mat-
      progress-bar>
  <img mat-card-avatar class="auth-user-image" src="{{
    authUser?.image }}" />
  <textarea #postText class="post-box"
    placeholder="What's on your mind, {{ userFirstName
      }}?"
   matInput></textarea>
  <button mat-button (click)="handlePostClick()"
    >Post</button>
  <button mat-button (click)="postImage.click()" >{{
    imageFile?.name?? 'Select Image..' }} </button>
  <input #postImage type="file" style="display:none"
    (change)="onFileSelected($event)" />
</mat-card-content>
</mat-card>
```

We use a card avatar to display the image of the authenticated user, a `textarea` element to enter the post text, and the `#postText` template reference to query the element from the template using the `ViewChild` decorator.

Next, we add two buttons; the first button's click event will be bound to the `handlePostClick()` method, which will submit the post data to the parent component. The second button will be used to trigger the click event of the hidden input field, which will be used to select an image from the user's computer.

When an image is selected, the change event is fired, and the `onFileSelected()` method is called, which sets the component's `imageFile` property.

5. Finally, we need to add some styles to the component. Open the `create-post.component.css` file and start by setting `box-sizing` to `border-box` for all elements in the template as follows:

```
* {
    box-sizing: border-box;
}
```

6. Set `margin` of the card element as follows:

```
mat-card {
    margin: 0.4rem;
}
```

7. Set `width` of the post box wrapper to `100%` as follows:

```
.post-box-wrapper {
    width: 100%;
}
```

8. Style the post box as follows:

```
.post-box {
    width: 90%;
    height: 50px;
    margin-left: 10px;
    padding: 10px;
    background-color: #f0f2f5;
}
```

9. Style the image of the authenticated user as follows:

```
.auth-user-image {
  background-color: gray;
  border-width: 1px;
  margin: 10px;
}
```

If the user does not upload a photo, we set a default background for the authenticated user image.

Let's use our dumb component in the profile component after we've created it:

10. Locate the `<div>` element with the `posts-column` class in the `users/components/profile/profile.component.html` file and add the component as follows:

```
<app-create-post [authUser]="authUser"
  [loading]="loading" (post)="onPost($event)"
*ngIf="isAuthUserProfile && authUser">
</app-create-post>
```

We use the selector defined in the component's decorator to invoke the component just like any other tag. We pass in the `authUser` and `loading` properties from the profile component, and we use Angular event binding syntax (that is, `()`) to bind the profile component's `onPost()` method to the `<app-create-post>` component's custom `post` event. As a result, the profile component can listen for events from the child component and access data passed through the `$event` object.

We use input properties and custom events to communicate between the parent and child components. We can pass data from the parent component to the child component using input properties, and we can pass data from the child component to the parent component using events.

This is how this component appears:

Figure 9.4 – The create post component

We've added a presentational component for creating text and image posts. This component displays a `<textarea>` element with the authenticated user's avatar and two buttons for selecting images and creating posts. A custom `post` event is used to delegate this final operation to the parent component. To be more specific, the `onPost()` method is called to create the post, which also delegated this process to the post service.

You can now manually test your app to see whether posting works properly. If you choose an image, enter some text, and then press the **Post** button, you should see a message stating that your post has been created. Go to your browser console, and you should see the created post displayed similar to the following screenshot:

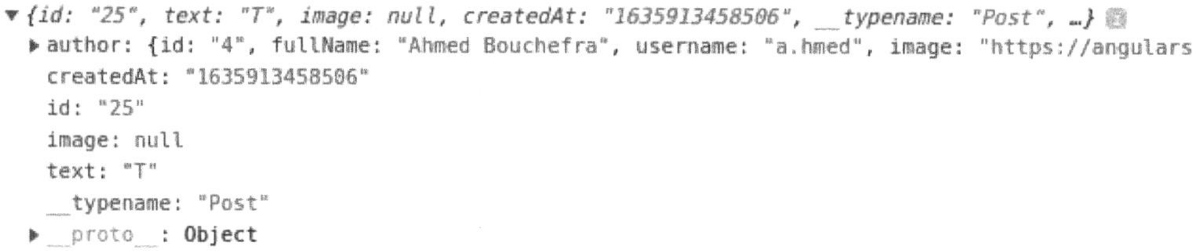

```
▼ {id: "25", text: "T", image: null, createdAt: "1635913458506", __typename: "Post", …} 📋
  ▶ author: {id: "4", fullName: "Ahmed Bouchefra", username: "a.hmed", image: "https://angulars
    createdAt: "1635913458506"
    id: "25"
    image: null
    text: "T"
    __typename: "Post"
  ▶ __proto__: Object
```

Figure 9.5 – Created post displayed on the console

The post with the nested `author` object is returned by the mutation. We can use Apollo Client to automatically update the posts count displayed on the profile section without manually updating it or refreshing the page because the nested `author` object is the current profile user who created the post.

As previously stated, if the mutation returns an object that already exists in the cache, Apollo Client will automatically select the modified fields and merge them with the existing object in the cache, after which the user interface will be automatically updated. However, in our case, this does not appear to work, but this is unrelated to the frontend. The updated author's posts count is not returned in the backend after creating a post. This can be confirmed using the browser console.

We can resolve this by opening the `server/src/graphql/resolvers.ts` file, searching for the post mutation, and then simply changing the order of the highlighted lines as follows:

```
post: async (_, args, { orm, authUser }: Context) => {
  const post = orm.postRepository.create(
    {
      text: args.text,
      image: args.image,
      author: await orm.userRepository.findOne(authUser?.id)
    } as unknown as PostEntity
```

```
  );
  await orm.userRepository.update({ id: authUser?.id }, {
    postsCount: post.author.postsCount + 1 });
  const savedPost = await orm.postRepository.save(post);
  return savedPost as unknown as Post;
}
```

This ensures that the author's postsCount field is updated before saving the post.

After adding a new post, our profile UI will now be automatically updated.

Testing

We'll need to add unit tests now that we've implemented the image uploading functionality, which includes setting the user's photo, cover image, and creating posts. As assignments, we'll test the post and profile services, followed by the profile component.

Write the following tests for the post service:

- should upload the file passed as argument
- should create a post
- should remove a post by ID

Then, write the following tests for the profile service:

- should set user bio
- should set user photo
- should set user profile cover

Fix the existing test and write more tests for the profile component:

- should invite unauthenticated users
- should display edit button for authenticated users
- should display the profile user information
- should set the profile user when found
- should set the auth user
- should display 'User not found!' if no profile user is found
- should hide edit button for other users profiles

- should render 'create post component' for authenticated user profiles
- should hide 'create post component' for other users profiles
- should pass the props to 'create post component'
- should handle the post event of 'create post component'
- should display edit section on button click
- should set bio on button trigger
- should set cover on input change
- should set photo on input change

Write the following tests for the create post component:

- should render progress bar when loading is true
- should render auth user image
- should render auth user first name on placeholder
- should emit post on button click
- should set imageFile property on file select
- should render selected image name

You can find the code for these assignments at https://git.io/JXRLn.

Summary

Throughout this chapter, we started implementing the profile component's functionality. We added the necessary code to fetch the user that corresponds to a profile URL and render their information on the page, including the ability to upload the user's photo and cover image, as well as adding a biography.

In the following chapter, we'll continue implementing our users' profile components before adding realtime subscriptions to notify users in real time if other network users liked or commented on their posts.

10
Fetching Posts and Adding Comments and Likes

In the previous chapters, we introduced you to Apollo Client and demonstrated how to integrate it with our Angular frontend. We also implemented authentication and image uploading, shielded the feed's posts component from unauthenticated users, and built a header component with the application's navigation buttons and a search bar to search for users in our network using their full names.

We then began creating a profile component, which allows users to add their biography, cover image, and photo, as well as create posts.

In this chapter, we'll begin by dealing with an authentication token expiration issue, which we'll describe later, and then we'll continue working on our profile component by sending queries to receive paginated posts and comments data, and mutations to add comments and likes to posts.

The following topics will be covered:

- Improving the authentication system
- Adding queries and generating types and services
- Fetching paginated posts
- Implementing comments and likes services
- Implementing a presentational `post` component
- Displaying posts, comments, and likes

Here's a summary of what we'll do:

1. First, we'll write query and mutation strings that will be sent to the server.
2. We'll utilize the GraphQL generator to produce the types of the results and variables, as well as the GraphQL services (that extend the `Query` and `Mutation` classes of `apollo-angular`) to like and comment on posts.
3. We'll create and export Angular services, which include the logic for working with comments and likes and sending requests to the server.
4. We'll add *posts-fetching-with-pagination* logic to the previous chapter's post service. We'll provide two methods for retrieving a user's posts, as well as a feed of the most recent posts made by application members.
5. We'll implement a `post` component, a dumb component for displaying a single post, with comments and likes, as well as buttons for liking and commenting on the post.
6. Finally, the posts, comments, and likes will be displayed on the profile and home views.

Technical requirements

To complete the steps in this chapter, you must first finish the previous chapter's steps and assignments and be familiar with the following technologies:

- JavaScript/TypeScript
- **HyperText Markup Language (HTML)**
- **Cascading Style Sheets (CSS)**

You can find the source code for this chapter at `https://github.com/ PacktPublishing/Full-Stack-App-Development-with-Angular-and- GraphQL/tree/main/Chapter08`.

Improving the authentication system

Before we can proceed with building the profile component, we must address a token expiration issue. When a token expires, users must return to the login page and sign in again to acquire a new token. Even after receiving the token and being referred to the user's profile page, the profile's posts are not retrieved, and users see a message indicating that they are not authenticated.

This is because, after re-authentication, the Apollo context is still utilizing the previous— expired—token, which we simply need to replace, as follows:

1. Import the following symbols into the `core/services/auth/auth.service.ts` file:

   ```
   import { createApollo } from 'src/app/graphql.module';
   import { HttpLink } from 'apollo-angular/http';
   ```

   The authentication token is obtained from local storage and supplied to the Apollo context in the `createApollo()` method.

   Apollo Client uses `HttpLink` to send GraphQL queries to the server. We are importing it since the `createApollo()` method requires it.

2. Inject `HttpLink` through the constructor, like this:

   ```
   export class AuthService {
     constructor(
       private httpLink: HttpLink) { /* ... */ }
   ```

   We inject `HttpLink` since it is required by the `createApollo()` method we'll be calling in the next step. See `https://apollo-angular.com/docs/data/network/#http-link` for more on this.

3. Locate the `login()` function and insert the following code just after the `updateAuthState()` call, within the `next` handler of the object provided to the `tap()` operator:

   ```
   const acOpts = createApollo(this.httpLink);
   this.apollo.client.setLink(acOpts.link!);
   ```

   We call the GraphQL module's `createApollo()` function, which causes the context link to obtain a new token from `localStorage`, and then we call the `setLink()` method to set the returned link chain and pass it to Apollo Client.

When the token expires after these changes, just log in again, and you should be able to run actions that need user authentication. It is strongly advised that you read the Apollo Link overview at `https://www.apollographql.com/docs/react/api/link/introduction/`.

Adding queries and generating types and services

In this section, we'll continue working on our profile component by sending queries to receive paginated posts and comments data, as well as mutations, to add comments and likes to posts.

Here is a screenshot of a single post prior to the addition of any comments or likes:

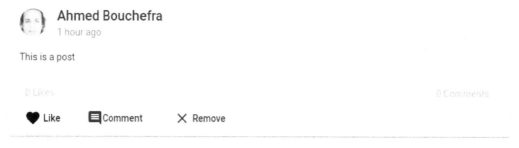

Figure 10.1 – A post with no comments or likes

Here is a screenshot of a post after I commented and liked it:

Figure 10.2 – A post after adding a comment and a like

Before we begin, let's change one behavior that we implemented in *Chapter 9, Uploading Images and Adding Posts*. We utilized the `takeUntil()` operator in the base component to unsubscribe from the returned Observables when the component is destroyed. This is the correct method when we need to keep Observables alive over the component's lifetime in order to keep watching active queries, but it is not necessary for the `createTextPost()`, `uploadImageAndCreatePost()`, and `onRemovePost()` methods. To release memory, it is preferable to utilize the `first()` or `take()` operators to unsubscribe immediately after receiving the first emitted value from the Observables.

As an assignment, import the `take()` operator from `rxjs/operators` and unsubscribe using the `take(1)` statement instead of `takeUntil(this.componentDestroyed)` in the mentioned methods.

Next, let's get things started by adding query and mutation strings that will be sent to the GraphQL server for getting posts, comments, and likes, as well as for adding comments and likes to existing posts. Proceed as follows:

1. Open the `graphql.documents/post.graphql` file and start by updating the `CommentsInfo` fragment, as follows:

```
fragment CommentsInfo on Post {
latestComment {
  ...CommentFields
  author {
    id fullName username image
  }
}
commentsCount
}
```

Rather than expanding the `BasicUserFields` fragment on the `author` field, we just ask for the required fields explicitly. This is only necessary for the time being to fix an unknown issue that causes a bad GraphQL request to be sent when fetching posts.

2. Add the following query to the same file:

```
query getPostsByUserId(
$userId: ID!,
$offset: Int,
$limit: Int) {
getPostsByUserId(
```

```
    userId: $userId,
    offset: $offset,
    limit: $limit) {
      ...PostFields
    }
  }
```

This is a GraphQL query that returns paginated posts based on the user **identifier** (ID). It accepts the user ID, offset, and limit as input. By expanding the `PostFields` fragment defined in *Chapter 9*, *Uploading Images and Adding Posts*, inside the selection set, we select all of the fields necessary for rendering the post.

3. Add the `getFeed` query to the same file to retrieve the user's feed, as follows:

```
query getFeed(
  $offset: Int,
  $limit: Int) {
  getFeed(
    offset: $offset,
    limit: $limit) {
      ...PostFields
    }
  }
```

This query is similar to the one before it. It is used to get all of the network's recent posts, not just those belonging to a certain user. It only accepts the offset and limit parameters necessary for paginating data as input. We expand the `PostFields` fragment in the selection set to ask for the same fields as the previous query.

4. Add the following query to the `graphql.documents/comment.graphql` file:

```
query getCommentsByPostId(
  $postId: ID!,
  $offset: Int,
  $limit: Int) {
  getCommentsByPostId(
    postId: $postId,
    offset: $offset,
    limit: $limit) {
      ...CommentFields
```

```
    author { ...BasicUserFields }
    post { id }
  }
}
```

This query document will be used to retrieve comments for a certain post. It accepts the post ID and the offset/limit arguments, for paginating comments, as input. In the selection set, we ask for the ID, comment text, and date, as well as sub-selections for author information (ID, full name, username, and image) and post (ID).

5. Add the following mutation to the same file to create a comment:

```
mutation commentPost(
  $comment: String!,
  $postId: ID!){
  comment(comment:$comment, postId:$postId){
    id post { id commentsCount }
  }
}
```

In this step, we define a GraphQL document that accepts the text of the comment and the post ID as input, both of which are required inputs. In the selection set, we select the ID of the new comment, the post's ID, and the commentsCount fields.

6. Add the following mutation to remove comments:

```
mutation removeComment($id: ID!){
  removeComment(id:$id){
    id post { id commentsCount }
  }
}
```

This mutation takes as an argument the ID of the comment to be removed.

7. Create a graphql.documents/like.graphql file and add the following query to it:

```
query getLikesByPostId(
  $postId: ID!,
  $offset: Int,
  $limit: Int){
  getLikesByPostId(
```

```
        postId:$postId,
        offset: $offset,
        limit: $limit){
        id
        user { ...BasicUserFields }
        post { id likesCount }
        createdAt
      }
    }
```

This query is used to retrieve a post's likes and accepts the post's ID as input as well as offset/limit parameters for paginating likes.

8. Add the following mutation to like a post:

```
mutation likePost(
  $postId: ID!){
  like(postId:$postId){
    id post {   id ...LikesInfo }
  }
}
```

This mutation takes the post's ID as input and requests the ID of the like that was saved in the server's database, as well as information about the related post.

9. Add the following mutation to remove likes by ID:

```
mutation removeLike(
  $postId: ID!){
  removeLike(postId:$postId){
    id post { id ...LikesInfo }
  }
}
```

This mutation takes the liked post's ID as input and requests the ID and post's details.

10. Return to your Terminal and execute the following command to invoke the code generator from the **monorepo** project's root:

```
npm run codegen
```

We added strings for the GraphQL queries and mutations that would be sent to the server and then we generated variables and results types as well as services for sending the requests.

Basically, we'll utilize a number of generated Angular services to fetch posts and add/remove comments or likes, which extend either the generic Query or Mutation classes from apollo-angular. By setting the document property to the appropriate query or mutation document and the response type within angle brackets, we define the shape of the result and variables.

To retrieve the user's posts, we'll use the Apollo service's generic watchQuery() method and set the type between brackets to the proper result and variable types.

Following that, we'll look at how to retrieve posts in the post service we built in the previous chapter.

Fetching paginated posts

In the post service, we'll define and implement a getPostsByUserId() method that will be called by the profile component to retrieve paginated posts belonging to a certain user. As an assignment, we'll also implement a getFeed() method to retrieve the network's most recent posts, regardless of who posted them. Proceed as follows:

1. Open the core/services/post/post.service.ts file and add the following imports to it:

    ```
    import { Apollo } from 'apollo-angular';
    import {
      GetPostsByUserIdDocument,
      GetPostsByUserIdQuery,
      GetPostsByUserIdQueryVariables
    } from '@ngsocial/graphql/documents';
    ```

    These imports were generated by the GraphQL code generator.

2. Inject the Apollo service, as follows:

    ```
    constructor(
      /* ... */
      private apollo: Apollo) { }
    ```

3. Define a `getPostsByUserId()` method, as follows:

```
getPostsByUserId(
  userId: string,
  offset?: number,
  limit?: number) {
}
```

This method takes the user ID as a required argument and the `offset/limit` pair as optional arguments.

4. Call the generic `watchQuery()` method of the injected `Apollo` service in the method body, passing the arguments as follows:

```
const queryRef = this.apollo
.watchQuery<
  GetPostsByUserIdQuery,
  GetPostsByUserIdQueryVariables>(/* ... */);
return queryRef;
```

The `watchQuery()` method returns a `QueryRef` object, but we're mostly interested in the `valueChanges` property, which is an RxJS Observable that we must subscribe to in order to receive the data, and the `fetchMore()` method, which we'll use to get more paginated data.

See `https://apollo-angular.com/docs/data/queries/#what-is-queryref` for more on this.

Then, pass the following object as an argument:

```
{
  query: GetPostsByUserIdDocument,
  variables: {
    userId: userId,
    offset: offset || 0,
    limit: limit || 10
  },
  fetchPolicy: 'cache-and-network',
}
```

We specify the query that will be sent to the GraphQL server, which, in this case, is the document for fetching the user's posts that we previously generated from the `getPostsByUserId` query.

Then, we provide the `variables` object with the arguments needed by the query, which are the `userId`, `offset`, and `limit` arguments.

Next, we set `fetchPolicy` to `cache-and-network` to instruct Apollo to check the cache first for available data and then send a request to the GraphQL server to get newer data and save it to the cache. This last step happens regardless of whether any data is found in the cache.

5. Open the `src/app/cache.ts` file and add the `offsetLimitPagination` import, as follows:

```
import { offsetLimitPagination } from
   '@apollo/client/utilities';
```

Next, update the cache by adding the following field policy for `Query`. `getPostsByUserId`:

```
getPostsByUserId: offsetLimitPagination(['userId'])
```

In the constructor of `InMemoryCache`, we provide the fields' policies to customize how the field is read and written in the cache of Apollo Client.

We use the `offsetLimitPagination` helper function to generate an appropriate field policy for the desired fields. This creates a `merge()` function for the field that can handle merging paginated results into the cache. This will allow us to use the `fetchMore` method to fetch more data.

Since the `getPostsByUserId` field accepts the `userId` argument as well as the `offset` and `limit` arguments, we need to set the user ID as a key argument for the field to enable Apollo Client to distinguish between posts that belong to each user. This is done by passing an array of key arguments to the `offsetLimitPagination` helper. See `https://www.apollographql.com/docs/react/pagination/offset-based/#setting-keyargs-with-offsetlimitpagination` for more on this.

For more information, see `https://apollo-angular.com/docs/data/pagination/`, and check out how to configure the cache at `https://apollo-angular.com/docs/caching/configuration/`.

After adding the `getPostsByUserId()` method for fetching posts by user ID and implementing pagination to the post service, we'll next see how to implement comments and likes services.

Implementing comments and likes services

Let's now implement services for fetching comments and likes. These two services, `CommentsService` and `LikesService`, encapsulate the logic for liking and commenting on posts, alongside methods for fetching posts' comments and likes with pagination. Proceed as follows:

1. Head back to your Terminal and run the following commands to generate `CommentsService` and `LikesService` instances:

```
ng g s core/services/comment/comments
ng g s core/services/like/likes
```

2. Open the `core/services/comment/comments.service.ts` file and start by adding the following imports to it:

```
import { StoreObject } from '@apollo/client/core';
import { map } from 'rxjs/operators';
import {
  CommentPostGQL,
  RemoveCommentGQL,
  GetCommentsByPostIdGQL
} from '../../gql.services';
```

Next, update the service constructor, as follows:

```
constructor(
  private commentPostGQL: CommentPostGQL,
  private removeCommentGQL: RemoveCommentGQL,
  private getCommentsByPostIdGQL:
    GetCommentsByPostIdGQL) { }
```

We simply inject the generated `CommentPostGQL`, `RemoveCommentGQL`, and `GetCommentsByPostIdGQL` services via the constructor. These are mutation and query services for creating, removing, and fetching comments that we've generated before.

3. Add the following method to the service to create comments:

```
createComment(
  comment: string,
  postId: string) {
  return this.commentPostGQL
```

```
    .mutate({
      comment: comment,
      postId: postId
    })
    .pipe(map(result => result.data!.comment));
}
```

We implement a method for creating comments that takes as arguments the comment and post ID and returns an Observable. In the body of the method, we call the `mutate()` method to send a mutation to create a comment. This method takes an object with attributes that correspond to the variables that are expected by the GraphQL mutation. The `mutate()` method returns an Observable, so we use the `pipe()` method and the `map()` operator to map the `result` object to the `data!.comment` object.

4. Add the following method to the service to remove comments by ID:

```
removeComment(id: string) {
  return this.removeCommentGQL
    .mutate({
      id: id
    })
    .pipe(map(result => result.data!.removeComment));
}
```

As with the previous method, we call the `mutate()` method of the mutation service to remove posts. We pass the `id` parameter, and we map the `result` object to the `data!.removeComment` object.

5. Add the following method to fetch comments by post ID:

```
getCommentsByPostId(
  postId: string,
  offset?: number,
  limit?: number) {
  const queryRef = this.getCommentsByPostIdGQL
    .watch(
      {
        postId: postId,
        offset: offset || 0,
        limit: limit || 5
```

```
        }, {
          fetchPolicy: 'cache-and-network'
        });
    return queryRef;
  }
```

Instead of calling the `mutate()` method, we call the `watch()` method for sending and watching the query to fetch comments from the server. We pass the variables required by the query, such as the post ID, offset, and limit, and we set the fetch policy to `cache-and-network`. Since the `offset` and `limit` arguments are optional, we also pass default values for pagination, which are zero for the offset and five for the limit.

The `Query` class provides two methods for fetching data—the `watch()` and `fetch()` methods, which are similar except that the `fetch()` method fetches data once. In our case, we need to use the `watch()` method since we need to implement pagination. Check out `https://apollo-angular.com/docs/data/services/#api-of-query` for more on this.

As an assignment, implement the following methods of `LikesService`:

- `likePost(postId: string)`
- `removeLike(postId: string)`
- `getLikesByPostId(postId: string, offset?: number, limit?: number)`

Start by importing the appropriate GraphQL services; next, inject them via the service constructor, and then call the methods for sending requests to the server.

The `getLikesByPostId()` method is similar to the `getCommentsByPostId()` method. You can implement it by doing one of the following:

- Injecting the `Apollo` service via the constructor and calling the generic `watchQuery()` method of the injected instance by specifying the appropriate query and variables that will be sent to the GraphQL server and the type parameter for the returned response's data

- Injecting the generated `GetLikesByPostIdGQL` mutation service and calling the `watch()` method

After implementing the necessary Angular services for communicating with the GraphQL server, we'll now see how to create a presentational (dumb) component for displaying each post with its likes and comments information.

Implementing a presentational post component

The post component will only be presentational. It doesn't inject any services and it communicates with the parent components (containers such as the feed posts and profile components) with input properties and events.

Check out https://www.webtutpro.com/smart-dumb-components-in-angular-3c51ae6efcc4 for more information about smart (container) and dumb (presentational) components.

The post component takes as input the post and authUser properties, which should be passed from the parent component(s) to enable the component to render the post and information about the post's author. It also emits a bunch of custom events to the parent components, such as the following:

- A comment event that fires when the user comments on a post
- A like event that fires when the user likes a post
- A remove event that fires when the user removes a post
- A listComments event that fires when the user wants to display a post's comments
- A moreComments event that fires when the user wants to display more posts' comments
- A listLikes event that fires when the user wants to display a post's likes

Let's get started with the steps to implement our post component:

1. Generate a post component using the following command:

```
ng g c shared/components/post
```

This component will be generated inside the post/ folder of the shared/components folder.

The component will be automatically imported into the shared module, but you also need to export it from the module by adding it to the exports array. This enables you to use the component in other modules where the shared module is imported.

2. Define event types; create a `shared/types/comment.event.ts` file and add the following types to it:

```
export type CommentEvent = {
  comment: string;
  postId: string;
};
```

Next, create a `removecomment.event.ts` file and add the following code to it:

```
export type RemoveCommentEvent = {
  id: string;
};
```

Next, create a `listcomments.event.ts` file and add the following code to it:

```
export type ListCommentsEvent = {
  postId: string;
};
```

Next, create a `morecomments.event.ts` file and add the following code to it:

```
export type MoreCommentsEvent = {
  postId: string;
};
```

These types define the shapes of data values that will be sent with the custom events.

Make sure to export these types from the `shared/index.ts` file, as follows:

```
export { CommentEvent }
    from './types/comment.event';
export { RemoveCommentEvent }
    from './types/removecomment.event';
export { ListCommentsEvent }
    from './types/listcomments.event';
export { MoreCommentsEvent }
    from './types/morecomments.event';
```

3. Open the `shared/types/like.event.ts` file and add the following type to it:

```
import { Post } from "@ngsocial/graphql/types";

export type LikeEvent = {
```

```
  post: Post;
};
```

Next, create a `removelike.event.ts` file and add the following code to it:

```
export type RemoveLikeEvent = {
  id: string;
};
```

Next, create a `displaylikes.event.ts` file and add the following code to it:

```
export type DisplayLikesEvent = {
  postId: string;
};
```

Also, export the following types from the `shared/index.ts` file:

```
export { LikeEvent }
    from './types/like.event';
export { RemoveLikeEvent }
    from './types/removelike.event';
export { DisplayLikesEvent }
    from './types/displaylikes.event';
```

4. Open the `shared/components/post/post.component.ts` file and start by adding these imports:

```
import {
  Input,
  Output,
  EventEmitter,
  ViewChild,
  ElementRef
} from '@angular/core';
```

Next, import the `User` and `Post` types, like this:

```
import {
  User,
  Post
} from '@ngsocial/graphql/types';
```

Finally, import the defined event types, like this:

```
import {
    RemovePostEvent,
    CommentEvent,
    ListCommentsEvent,
    MoreCommentsEvent,
    LikeEvent,
    DisplayLikesEvent
} from '../..';
```

These will be imported from the barrel file—the `index.ts` file—of the `shared/` folder.

5. Define the input properties, as follows:

```
@Input() post!: Post;
@Input() authUser: Partial<User> | null = null;
@Input() commentsPerPost: number = 5;
```

We define three properties—`post`, `authUser`, and `commentsPerPost`—and decorate them with the `@Input` decorator to enable the `post` component to receive their values from its parent component(s) (the profile and feed's posts components).

We use the non-null assertion operator (`!`) to assert that the post input will not be `null`. Check out `https://www.typescriptlang.org/docs/handbook/release-notes/typescript-2-0.html#non-null-assertion-operator` for more on this.

6. Define the output events, as follows:

```
@Output()
like = new EventEmitter<LikeEvent>();
@Output()
comment = new EventEmitter<CommentEvent>();
@Output()
listComments = new EventEmitter<ListCommentsEvent>();
@Output()
moreComments = new EventEmitter<MoreCommentsEvent>();
```

```
@Output()
remove = new EventEmitter<RemovePostEvent>();
@Output()
listLikes = new EventEmitter<DisplayLikesEvent>();
```

We define output properties and decorate them with the `@Output` decorator; then, we assign an instance of `EventEmitter` to each one and set the type parameter of the generic `EventEmitter` class to the appropriate custom event type that we defined earlier. The type passed corresponds to the type of object that will be sent to the parent components.

7. Define the properties, as follows:

```
@ViewChild('commentInput')
commentInput!: ElementRef;
commentsShown: boolean = false;
```

We define a `commentInput` property of the type `ElementRef` and decorate it with the `@ViewChild` decorator so that we can access the element with the `commentInput` reference in the associated template. Then, we define a `commentsShow` property for controlling whether comments are displayed.

8. Add the following getter, which simply returns the number of likes concatenated with a `Likes` string:

```
get latestLike(): string {
    return '${this.post?.likesCount ?? 0} Likes';
}
```

Next, define the stub methods, as follows:

```
displayLikes(): void {}
displayComments(): void {}
loadComments(): void {}
sendLike(): void {}
removePost(): void {}
createComment(e: Event): void {}
```

The code for these methods will be added later, but for now, they are required in order to display the component without errors.

9. Implement a template for the component. Open the `shared/components/` `post/post.component.html` file and start by adding a Material card to display a post, as follows:

```
<mat-card class="post-wrapper" *ngIf="post">
</mat-card>
```

This card will be conditionally rendered only if the `post` input is not null. Since we've used the non-null assertion operator with the `post` input that will be passed from the `*ngFor` directive in the parent component, you can safely get rid of conditional rendering in this case, if you prefer.

10. Inside the card, add a header, as follows:

```
<mat-card-header>
  <img mat-card-avatar class="post-user-avatar"
    [src]="post.author?.image" />
  <mat-card-title>
    {{post.author?.fullName}}
  </mat-card-title>
  <mat-card-subtitle>
    {{ post.createdAt }}
  </mat-card-subtitle>
</mat-card-header>
```

Next, add a content section, as follows:

```
<mat-card-content>
  <p>
    {{post.text}}
  </p>
  <img class="post-card-image" mat-card-image
    *ngIf="post.image" [src]="post.image"
    [alt]="post.text">
</mat-card-content>
```

Next, add an actions section for the card, as follows:

```
<mat-card-actions>
</mat-card-actions>
```

We use a Material card for displaying the post using information such as the full name of the post's author, the date on which a post is created, and the post's text and image. We use Angular interpolation to display the value of the `author.` `fullName`, `createdAt`, and `text` properties, and we use property binding to bind the `src` attribute of the `<image>` element to the `post.image` property.

The image is conditionally rendered using the `*ngIf` directive only if the `post.` `image` property is not null.

11. Inside the card's actions section, add the following HTML division with a `class` attribute that has a value of `engagments-count`:

```
<div class="engagements-count">
  <button [disabled]="post.likesCount === 0" mat-
    button (click)="displayLikes()">
  {{ latestLike }}
  </button>
  <button [disabled]="post.commentsCount === 0" mat-
    button (click)="displayComments()">
  {{ post.commentsCount }} Comments
  </button>
</div>
```

We add two buttons that display the number of likes and comments, then we bind their `click` events with the `displayLikes()` and `displayComments()` methods to send events to display likes and comments. We display the likes number at the left and the comments number at the right. We disable the buttons if there are no likes or comments.

12. Add a second HTML division with a `class` attribute that has a value of `engagement-actions`, as follows:

```
<div class="engagement-actions">
  <button [style.color]="post?.likedByAuthUser ? 'red'
    : 'black'" mat-button (click)="sendLike()">
  <mat-icon>favorite</mat-icon>
  <span>
  Like
  </span>
  </button>
  <button mat-button (click)="displayComments()">
```

```
        <mat-icon>comment</mat-icon>
        <span>Comment</span>
    </button>
    <button *ngIf="authUser?.id == post.author?.id" mat-
        button (click)="removePost()">
    <mat-icon aria-hidden="false" aria-
        label="Remove">clear</mat-icon>
    <span>
    Remove
    </span>
    </button>
  </div>
```

We add three Material buttons with icons and bind their `click` event to the `sendLike()`, `displayComments()`, and `removePost()` methods. These methods are used to send likes, add comments, and remove posts. The **Remove** button is conditionally rendered using the `*ngIf` directive so that it is shown only if the post's author is the authenticated user.

The safe navigation operator (?) was used here since the `authUser` object may be null.

13. Add a third division with a `class` attribute containing a `comment-area` value, as follows:

```
    <div class="comment-area"></div>
```

Inside the division, add a card for displaying the latest comment, as follows:

```
    <mat-card *ngIf="!commentsShown && post.latestComment"
      class="comment-card">
    <mat-card-content class="card-comment-content">
        <h4>{{post.latestComment?.author?.fullName}}</h4>
        {{post.latestComment?.comment}}
    </mat-card-content>
    {{ post.latestComment?.createdAt }}
    </mat-card>
```

We conditionally render a Material card for displaying the latest comment if the post has a latest comment and the comments' feed is not shown.

Then, add a division for submitting comments, as follows:

```
<div class="comment-box" *ngIf="commentsShown">
  <textarea #commentInput
  placeholder="Leave a comment.."
  matInput
  (keydown.enter)="createComment($event)">
  </textarea>
</div>
```

We conditionally render this HTML division, which contains a `<textarea>` element, referenced with a #commentInput reference, for entering comments. We bind the keydown.enter event of the `<textarea>` element to the createComment() method to allow users to create comments by pressing the *Enter* key on the keyboard.

Finally, add a button for loading more comments, as follows:

```
<button mat-button class="load-comments-btn"
  (click)="loadComments()"
 *ngIf="commentsShown && post.commentsCount >
  commentsPerPost">
Load more comments
</button>
```

This displays a Material button with the click event bound to the loadComments() method for loading more comments.

14. Open the post.component.css file and start by adding the following styles to it:

```
* {
  box-sizing: border-box;
}
mat-card {
  margin: 0.4rem;
}
.post-wrapper {
  width: 100%;
}
.post-card-image {
  margin:0px;
```

```
    max-width: 100%;
    max-height: 100%;
    width: auto;
    height: auto;
}
```

15. Add the following class for styling the post's user avatar:

```
.post-user-avatar {
    background-color: gray;
    border-width: 1px;
}
```

16. Style the division with the `engagements-count` class using the following styles:

```
.engagements-count {
    display: flex;
    flex-direction: row;
    width: 100%;
    justify-content: space-between;
}
```

17. Style the comment elements using the following styles:

```
.comment-box > textarea {
    width: 100%;
    padding: 10px;
    background-color: #f0f2f5;
    border-radius: 9px;
}
```

We have implemented a `post` component for displaying a single post with its comments and likes information. Let's continue in the next section by fetching posts using the post service and displaying each one of them using this dumb component.

Displaying posts, comments, and likes

After partially implementing our presentational `post` component, let's see how to use it in the parent user's profile component and use its input properties and custom events to communicate between the child and parent. Here are the steps we need to follow:

1. First, we need to include the `post` component in the profile component, so open the `users/components/profile/profile.component.html` file and add the following markup in the `<div>` element, with the `posts-column` class, below the `<app-create-post>` component:

    ```
    <div *ngFor="let post of posts;">
      <app-post [post]="post" [authUser]="authUser"
      (remove)="onRemovePost($event)">
      </app-post>
    </div>
    ```

    We include a component in the template using the value of the `selector` attribute in the component's decorator, which is `app-post` in this case.

    We can invoke this component without issues since it's exported from the shared module, which is, by itself, imported from the `users` module that contains the profile component.

    We iterate over the `posts` array using the `*ngFor` directive and display each post using the `post` component. We pass in the `post` object via the `post` property of the component and the authenticated user using the `authUser` property. Then, we bind the custom `remove` event of the dumb component to the `onRemovePost()` method, which is implemented and inherited from the base component.

    The component accepts other custom events, but we first need to define and implement handlers for these in the base class before binding them here. We add these handler methods to the base class to be able to reuse them from the `post` component of the feeds module.

2. Add the code for fetching the posts to the profile component's class. Open the `users/components/profile/profile.component.ts` file and start by importing the `Post` type by adding it to the existing import declaration, as follows:

    ```
    import { Post, User }
      from '@ngsocial/graphql/types';
    ```

Then, update the existing code inside the `ngOnInit()` method by calling the `getPostsByUserId()` method of the post service inside a second call to the `switchMap()` operator, as follows:

```
switchMap((userResponse) => {
  this.setProfileUser(userResponse.getUser);
  this.setIsAuthUserProfile();
  return this.postService
    .getPostsByUserId(userResponse.getUser.id)
    .valueChanges;
})
```

This will flatten the `getUser` Observable returned from the previous call of `switchMap()` to the `getPostsByUserId` Observable.

We also moved the call to the `setProfileUser()` and `setIsAuthUserProfile()` methods from the `subscribe()` method of the `userObs` Observable to the body of the second `switchMap()` call since this is where we receive the `getUser` object containing the profile's user information.

The `getPostByUserId()` method returns a `QueryRef` object that contains properties such as the `valueChanges` Observable.

After that, you need to set the `loading` property to `true` and update the `next` handler of the object passed to the `subscribe()` method, as follows:

```
userObs
.pipe(getPostsObs => {
  this.loading = true;
  return getPostsObs;
})
.subscribe({
  next: (result) => {
    this.loading = result.loading;
    this.posts = result
      .data
      .getPostsByUserId as Post[];
    console.log(this.posts);
  },
  error: (err) => super.handleErrors(err)
});
```

In the case of success, we get an `ApolloQueryResult` object that contains the received data and the `loading` property, indicating whetherthe query is still loading.

The Observable will emit when the query is complete, and `loading` will be set to `false`. The query also has a `data` object with a `getPostsByUserId` member, which is the field that we asked for in the corresponding GraphQL operation. See `https://apollo-angular.com/docs/data/queries/` for more on this.

At this point, you should be able to fetch and display each post using the presentational component. Additionally, if you create a post, it will be automatically displayed in the posts' list without refreshing the page or adding any extra code, thanks to Apollo Client and its cache.

Here is a screenshot of a post:

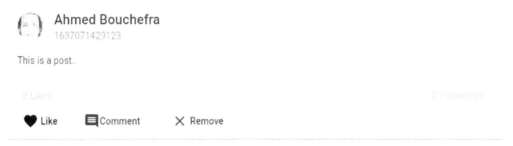

Figure 10.3 – A single post

You can see that the date of when the post is added is not displayed in a human-readable format. We'll fix that later using field policies that control how fields are read and/or written.

If you click on the buttons, none of them work. We've already implemented an `onRemovePost()` method for removing a post by ID in the previous chapter and assigned a method to the custom `remove` event of the `<app-post>` element, but the button is still not working. This is simply because when clicking on the button, an event is not emitted.

3. Let's emit an appropriate event when the **Remove** button is clicked. Go back to the `shared/components/post/post.component.ts` file and update the `removePost()` method, as follows:

```
removePost(): void {
  this.remove.emit({
    id: this.post.id
  });
}
```

We call the emit() method of the custom remove event to emit an event to the parent component(s) and we pass the ID of the post that should be removed as the event's value.

Now, if you click on the button, the post should be deleted. You can confirm that by refreshing the component since Apollo Client has no way to automatically remove a post from the cache after it has been removed from the server.

We can use various techniques to remove a post from the cache either by manually updating the cache or refetching related queries, such as getUser and getPostsByUserId.

Check out https://apollo-angular.com/docs/caching/advanced-topics and https://www.apollographql.com/docs/react/data/refetching/ for more on this.

Since we've previously seen how to update the cache manually when implementing the comments service, let's use query refetching in this method.

4. Open the core/services/post/post.service.ts file and update the removePost() method, as follows:

```
removePost(id: string) {
  return this.removePostGQL
    .mutate({
      id: id
    }, { refetchQueries: ['getUser'] })
    .pipe(map(result => result.data!.removePost));
}
```

We use the refetchQueries option of the second object passed to the mutate() method, which takes an array of the queries you want to refetch. Here, we pass the getUser query used to fetch the profile's user, which will also cause the getPostsByUserId query to refetch since their Observables are chained using the switchMap() operator.

We are using this option only for learning purposes, but it may not be the best approach for performance reasons since it will send an extra request to the server.

As an assignment, implement other methods of the post component for emitting appropriate custom events.

5. For the `createComment()` method, you need to get the value entered in the `<textarea>` element, emit the custom `comment` event, and then clear the `<textarea>` element, as follows:

```
createComment(e: Event): void {
  e.preventDefault();
  this.comment.emit({
    comment: this.commentInput.nativeElement.value,
    postId: this.post.id
  });
  if (this.commentInput) {
    this.commentInput.nativeElement.value = '';
  }
}
```

We call the `emit()` method of `EventEmitter` to emit the custom event to the parent components with the appropriate values. We emit the `comment` event with an object containing the comment's text and the post ID as the event's value that will be passed to the parent component. Finally, we clear the comment input using the `nativeElement.value` properties of the `commentInput` property that is bound to the comment's input element in the template.

6. For the `displayComments()` method, you need to switch the value of the `commentsShown` property, which controls whether comments are displayed or not, and then emit the `listComments` event only if the value is `true`. Here's how to do this:

```
displayComments(): void {
  this.commentsShown = !this.commentsShown;
  if (this.commentsShown) {
    this.listComments.emit({
      postId: this.post.id
    });
  }
}
```

This method is bound to the `click` event of two buttons—the button that displays the number of the post's comments, and the **Comment** button.

In the `sendLike()` method, we emit the custom `like` event with the `post` object as a value.

After implementing the method, if you click on one of these buttons—at least, on one for now—the other button will be disabled since your post should initially have no comments, and you should see a comment box displayed, as in the following screenshot:

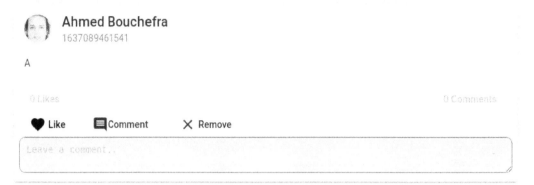

Figure 10.4 – A single post with a comment box displayed

Next, let's add code for fetching paginated posts.

7. Go back to the `users/components/profile/profile.component.ts` file and add the following property to the component:

```
fetchMore!: () => void;
```

This will be used to reference the method for fetching more posts, available from the `QueryRef` object returned from the call to the `getPostsByUserId()` method of the post service in the `ngOnInit()` method of the component.

8. Update the code for fetching posts in the `ngOnInit()` method, as follows:

```
switchMap((userResponse) => {
    this.setProfileUser(userResponse.getUser);
    this.setIsAuthUserProfile();
    const qRef = this.postService
        .getPostsByUserId(userResponse.getUser.id);
    this.fetchMore = () => {
        qRef.fetchMore({
            variables: {
                offset: this.posts.length
            }
        });
    }
}
```

```
    return qRef.valueChanges;
})
```

We assign the returned `QueryRef` object to the `qRef` constant, and then we assign an arrow function to the `fetchMore` property. This function simply wraps the `qRef.fetchMore()` method to set the `offset` variable to the `posts.length` value so that we only fetch posts that aren't already fetched.

Then, we return the `valueChanges` Observable, as in the previous step.

9. Add a button for fetching posts to the profile's template, as follows:

```
<button *ngIf="authUser && profileUser.postsCount &&
profileUser.postsCount > posts.length" mat-raised-button
color="primary" (click)="fetchMore()">More</button>
```

We simply bind the `click` event of this button to the `fetchMore()` function available as a property on the component's class.

This button will be conditionally rendered only if the user is logged in and the fetched posts' length, `posts.length`, is less than the number of posts that belong to the profile's user retrieved from the `profileUser.postsCount` property. We use the `*ngIf` directive to conditionally render this button. We use the `mat-raised-button` directive to add Material styles and behavior to the button, and we give it a primary color.

> **Note**
>
> The `fetchMore()` method is provided by Apollo for data pagination. Check out `https://apollo-angular.com/docs/caching/interaction#incremental-loading-fetchmore` for more information.

Let's move on to implementing code for adding comments and likes. When the user enters a comment and presses *Enter*, the `createComment()` method of the `post` component will be invoked.

We've already implemented this method, which emits the custom `comment` event to the parent profile component alongside required information, such as the comment itself and the post ID, and clears the comment box.

Similarly, when the user clicks on the **Like** button, displaying the black heart, the `sendLike()` method will be invoked and will emit the custom `like` event alongside the post that should be liked.

Since the `post` component is dumb, its role consists of emitting events to the parent component that delegates tasks for creating comments and likes to the appropriate services.

Let's implement the handlers of both these events in the base component and bind them on the `<app-post>` element. Proceed as follows:

1. Open the `core/components/base.component.ts` file and start by importing both `CommentsService` and `LikesService` and the results' types, and then inject the services, as follows:

```
import { CommentsService }
  from '../services/comment/comments.service';
import { LikesService }
  from '../services/like/likes.service';
```

Then, define the services' instances, as follows:

```
public commentsService: CommentsService;
public likesService: LikesService;
```

Finally, inject them inside the constructor, like this:

```
this.commentsService = injector.get(CommentsService);
this.likesService = injector.get(LikesService);
```

We use the `get()` method of Angular's `injector` instance (which is injected via the constructor) to inject `CommentsService` and `LikesService`, and then we assign the instances to the `commentsService` and `likesService` properties.

You also need to add the following highlighted types to the existing `import` declaration:

```
import {
  CommentEvent,
  LikeEvent,
  PostEvent,
  RemovePostEvent
} from 'src/app/shared';
```

2. Add the following method for handling the `comment` event:

```
onComment(e: CommentEvent): void {
  this.commentsService
    .createComment(e.comment, e.postId)
```

```
      .pipe(take(1))
      .subscribe({
        error: (err) => this.handleErrors(err)
      });
  }
```

When the `comment` event is emitted from the child `post` component, this method will be used to handle the event. It takes as an argument an object of the custom `CommentEvent` type.

Inside the method, we call the `createComment()` method of `CommentService`, and we pass the comment and post ID arguments that we get from the `event` object emitted from the child component. We subscribe to the returned Observable, and then we pass an `observer` object with an `error` handler that simply invokes and passes the `error` object to the `handleErrors()` method.

We use the `take()` operator to get the emitted value and unsubscribe from the returned Observable.

3. Implement an `onLike()` method for handling the custom `like` event, like this:

```
onLike(e: LikeEvent): void {
  if (e.post.likedByAuthUser) {
    this.likesService
      .removeLike(e.post.id)
      .pipe(take(1))
      .subscribe({
        error: (err) => this.handleErrors(err)
      });
  } else {
    this.likesService
      .likePost(e.post.id)
      .pipe(take(1))
      .subscribe({
        error: (err) => this.handleErrors(err)
      });
  }
}
```

When the custom `like` event is emitted from the child component, we'll invoke this method to handle it. If the `likedByAuthUser` property of the target `post` object is `true`, which means that the post is already liked by the authenticated user, we call and subscribe to the `removeLike()` method to unlike the post. Otherwise, we call and subscribe to the `likePost()` method to like that post. In both cases, we use the `take()` operator to unsubscribe from the returned Observable after emitting the first value.

4. Go back to the profile component's template and bind the defined methods to the `comment` and `like` events of the `<app-post>` element, as follows:

```
<app-post [post]="post" [authUser]="authUser"
    (remove)="onRemovePost($event)"
    (comment)= "onComment($event)"
    (like)= "onLike($event)">
</app-post>
```

Following the implementation of these methods, we will be able to like and comment on posts. Go to your app, like and comment on a post, and you should see the following **user interface (UI)** changes:

*   The displayed number of likes and comments will be incremented by one and the buttons will be enabled.

*   The **Like** button will change color from black to red, which denotes that the post is liked by the authenticated user. If you click on the button again, the post will be unliked and the color will change back to black.

*   The added comment will be displayed as the latest comment just below the **Post** button.

Since we asked the server, in the `commentPost` and `likePost` mutations, to return the ID along with `commentsCount` of the updated `post` object after we create a comment, and the `likedByAuthUser` attribute along with the `likesCount` attribute of the post, Apollo Client will pick up on these changes and automatically merge them with the existing post (identified by the ID) in the cache. This will rerun the active `getPostsByUserId` query and will cause the related Observable, returned from the `watchQuery()` method, to re-emit the posts data, which will automatically update the UI thanks to the change-detection and data-binding mechanisms of Angular—the `ngFor` directive will re-render the `<app-post>` element since its `post` input changes.

This is all good magic for us! Except for one side effect—the post element will get initialized, and properties such as commentsShown will have their initial false value, which will hide the comment box (and the comments' feed later) and display the latest comment.

Here is a screenshot of a post after liking and commenting on it:

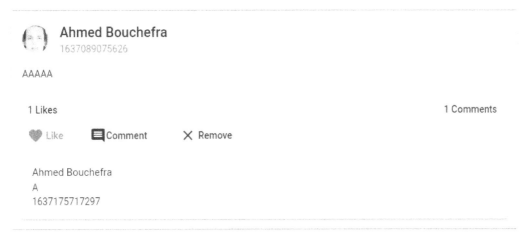

Figure 10.5 – A post after adding a comment and a like

This is not the desired behavior since we don't want the component to reset its internal state every time a user comments on or likes a post. Instead, the user should be presented with the comment box and the comments' feed (that will be implemented later) until they choose to hide them. We'll see how to implement these in the next chapter.

Summary

We've shown how to enhance our authentication system to deal with token expiration and how to utilize Apollo Client to send queries and mutations to our previously developed backend **application programming interface** (**API**). We've continued working on the profile component, adding code for getting paginated posts and comments, as well as making comments and likes.

In the following chapter, we'll finish up building the profile component before learning how to add realtime support to our application so that we can retrieve and display new data from the server without having to constantly refresh the app.

Part 3: Adding Realtime Support

In this part, we'll continue building our users' profile components before learning how to add realtime support to our application so that we can retrieve and display new data from the server without having to constantly refresh the app. We'll utilize GraphQL subscriptions with Apollo Client and Angular to do this.

This section comprises the following chapter:

- *Chapter 11, Implementing GraphQL Subscriptions*

11
Implementing GraphQL Subscriptions

In the previous chapter, we learned how to use Apollo Client to send queries and mutations to the backend API that we implemented previously to fetch paginated posts and comments, as well as create new comments and likes.

In this chapter, we'll continue building our users' profile components before learning how to add realtime support to our application. This will allow us to retrieve and display new data from the server without having to constantly refresh the app. We'll utilize GraphQL subscriptions with Apollo Client and Angular to do this.

Then, we'll configure Apollo Client for GraphQL subscriptions and show you how to get and display notifications in the application's header through a badge. GraphQL subscriptions allow the app to acquire new data from the server without having to manually refresh the app.

In this chapter, we will cover the following topics:

- Persisting the component state with reactive variables
- Using field policies to rewrite dates

- Displaying comments
- Setting up Apollo Client for subscriptions
- Implementing GraphQL subscriptions

Technical requirements

To complete this chapter, you must have Node.js and npm installed on your local development machine. Please refer to *Chapter 1*, *App Architecture and Development Environment*, to learn how to install them if you haven't done so already.

You also need to have completed the steps and assignments that were mentioned in the previous chapter.

Finally, you need to be familiar with the following technologies:

- JavaScript/TypeScript
- HTML
- CSS and SCSS

You can find the source code for this chapter at `https://github.com/ PacktPublishing/Full-Stack-App-Development-with-Angular-and-GraphQL/tree/main/Chapter08`.

Persisting the component state with reactive variables

In the previous chapter, we implemented some methods that allow us to like and comment on posts. However, there's one issue with this – after commenting or liking the post, the post element will be initialized and properties such as `commentsShown` will have an initial value of `false`, which will hide the comment box (and the comments' feed later) and display the latest comment.

This is not quite the intended behavior since we don't want the component to reset its internal state each time the user comments or likes a post. Instead, the user should be presented with the comment box and the comments' feed (which will be implemented later) until they choose to hide them.

The following screenshot shows a post once it's been liked and commented on:

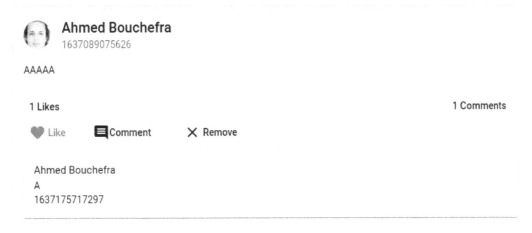

Figure 11.1 – A post once it's been liked and commented on

This feature's implementation is a good candidate for using state management with Apollo Client. The idea is to use the cache or a reactive variable to save the component's state, such as the commentsShown property value, which we need to persist when the component gets re-rendered.

Let's learn how to use a reactive variable since this is a more straightforward method:

1. Open the src/app/reactive.ts file and define and export the reactive variable for storing a Boolean value:

    ```
    export const commentsShownState =
      makeVar<boolean>(false);
    ```

2. Open the shared/components/post/post.component.ts file and import the defined reactive variable:

    ```
    import { commentsShownState }
      from 'src/app/reactive';
    ```

    Update the displayComments() method to set the value of the reactive variable to the value of the commentsShown property:

    ```
    displayComments(): void {
      this.commentsShown = !this.commentsShown;
      if (this.commentsShown) {
        this.listComments.emit({
          postId: this.post.id
        });
    ```

```
    }
    commentsShownState(this.commentsShown);
}
```

3. Finally, once the component has been initialized, set the value of the `commentsShown` property to the value of the reactive variable:

```
ngOnInit(): void {
    this.commentsShown = commentsShownState();
}
```

After implementing this, the `commentsShown` state of the component will be preserved. Now, if you display the comment box and add a comment or like to the post, it will still be displayed. This can be seen in the following screenshot, which was taken right after adding a comment:

Figure 11.2 – A post with a preserved state after adding a comment

Now, if you like the post, the state will also be preserved.

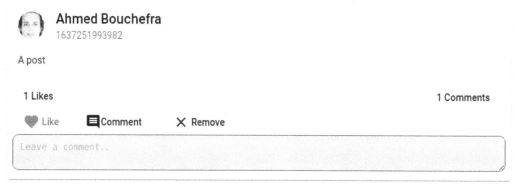

Figure 11.3 – A post with a preserved state after adding a like

If more than five comments are added, the **Load more comments** button will be displayed:

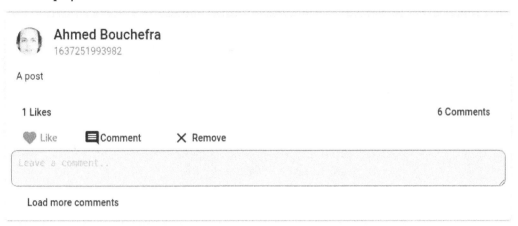

Figure 11.4 – A post with the "Load more comments" button

Then, if you click on the **Comment** button or the button that is displaying the number of comments, `<textarea>` and the **Load more comments** button will be hidden and the latest comment will be shown instead.

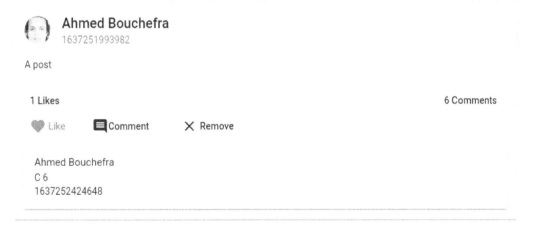

Figure 11.5 – A post with the latest comment displayed

When you add a post, Apollo Client doesn't always show it, so we need to refresh the app to see it. Let's make sure it always gets added without us having to refresh. Similar to removing posts, we can either re-fetch the related queries or manually update the cache. Open the `src/app/core/services/post/post.service.ts` file and update the `createPost()` method, as follows:

```
createPost(text: string | null, image: string | null) {
  return this.createPostGQL
    .mutate({
      text: text,
      image: image
    }, { refetchQueries: ['getUser']})
    .pipe(map(result => result.data!.post));
}
```

Here, we re-fetch the active `getUser` query, which will cause the related Observable to re-emit the updated profile's user object (the `postsCount` property has been incremented). This will also cause the `getPostsByUserId` Observable to re-emit the posts data since they are both chained by the `switchMerge()` operator.

Using field policies to rewrite dates

Next, let's display the dates of both the post and comments in a human-readable format. We will do this in the *time ago* format, which is commonly used in social network apps:

1. Install the following package in your client's project:

    ```
    npm install timeago.js
    ```

2. Open the `src/app/cache.ts` file and import the library:

    ```
    import * as timeago from 'timeago.js';
    ```

3. Add two field policies for the `Post.createdAt` and `Comment.createdAt` fields. Let's start with the `Post` type:

    ```
    Post: {
      fields: {
        createdAt: {
          read(createdAt) {
            return timeago.format(createdAt)
          }
    ```

```
            }
        }
    },
```

We simply return the formatted `createdAt` date when the field is read using the `format()` method of the `timeago` object we just imported. You can learn more at `https://timeago.org/`.

Then, add a similar field policy to the `Comment` type:

```
Comment: {
    fields: {
        createdAt: {
            read(createdAt) {
                return timeago.format(createdAt)
            }
        }
    }
}
```

You can find out more about the `read()` function at `https://www.apollographql.com/docs/react/caching/cache-field-behavior/#the-read-function`.

The following screenshot shows a post and a comment with their dates formatted using the *time ago* format:

Figure 11.6 – A post and comment with formatted dates

Now that we've formatted the post and latest comment dates using the "time ago" format, let's learn how to fetch and display comments.

Displaying comments

Follow these steps to fetch and display comments:

1. In the `core/components/base.component.ts` file, import the `Comment` type by adding it to the existing import declaration:

    ```
    import {
      User,
      Post,
      Comment
    } from '@ngsocial/graphql/types';
    ```

    Then, import the following highlighted types:

    ```
    import {
      CommentEvent,
      LikeEvent,
      ListCommentsEvent, MoreCommentsEvent,
      PostEvent,
      RemoveCommentEvent,
      RemovePostEvent
    } from 'src/app/shared';
    ```

2. In the same file, inside the base component, define the following property:

    ```
    public comments: Map<string,
      {
        result: Comment[],
        fetchMore: () => void
      }> = new Map();
    ```

    This will be used to map the `comments` array and the `fetchMore()` function to the posts IDs, which will allow us to save and track the comments and pagination function of each post.

3. Add the `onListComments()` method for fetching comments to the base class:

    ```
    onListComments(e: ListCommentsEvent): void {}
    ```

Inside the body of the method, call the getCommentsByPostId() method of commentsService:

```
const qRef = this.commentsService
  .getCommentsByPostId(e.postId);
```

Define the fetchMore() function, which simply wraps the fetchMore() method of the returned QueryRef object:

```
const fetchMore = () => {
  qRef.fetchMore({
    variables: {
      offset: this.comments
        .get(e.postId)?.result.length
    }
  })
}
```

Subscribe to the valueChanges Observable of the QueryRef object:

```
qRef
  .valueChanges
  .pipe(takeUntil(this.componentDestroyed))
  .subscribe();
```

Pass the following observer object to the subscribe() method:

```
{
  next: (result) => {
    this.comments
      .set(
        e.postId,
        {
          result: result
            .data
            .getCommentsByPostId as Comment[],
          fetchMore: fetchMore
        });
  }
}
```

This method is used to fetch a post's comments by ID. It takes the custom event object as an argument, calls the `getCommentsByPostId()` method of `CommentsService`, and subscribes to the `valueChanges` Observable after applying the `takeUntil()` operator with the `componentDestroyed` subject to unsubscribe when the component is destroyed.

In the `next()` handler, we simply put the fetched comments alongside the `fetchMore()` function in the `comments` map with the post ID, which the comments belong to, as the key. This will enable us to save this information next time we re-call the method to fetch the comments of another post.

4. Add the `onLoadMoreComments()` method, for fetching more comments, to the base class:

```
onLoadMoreComments(e: MoreCommentsEvent): void {
    const fetchMore = this.comments
      .get(e.postId)?.fetchMore!;
    fetchMore();
}
```

Add the `onRemoveComment()` method to remove comments by ID:

```
onRemoveComment(e: RemoveCommentEvent) {
    this.commentsService
      .removeComment(e.id)
      .pipe(take(1))
      .subscribe({
        error: (err) => this.handleErrors(err)
      });
}
```

5. Open the `src/app/cache.ts` file and add a policy for the `getCommentsByPostId` field using the `offsetLimitPagination` helper:

```
getCommentsByPostId: offsetLimitPagination(['id',
  'postId'])
```

6. Generate a dumb component for representing a single comment:

```
ng g component shared/components/comment
```

Make sure that you export the component from the shared module by adding it to the `exports` array.

7. Open the `shared/components/comment/comment.component.ts` file and add the following imports:

```
import {
  Component,
  EventEmitter,
  Input,
  OnInit,
  Output
} from '@angular/core';
import { RemoveCommentEvent }
  from '../../types/removecomment.event';
import { Comment }
  from '@ngsocial/graphql/types';
```

8. Add the following input and output properties to the component:

```
@Input() comment: Comment | null = null;
@Output() removeComment:
  EventEmitter<RemoveCommentEvent>
    = new EventEmitter();
```

9. Add the `remove()` method to the component that emits the custom removeComment event:

```
remove() {
  this.removeComment.emit({
    id: this.comment?.id ?? ''
  })
}
```

10. Open the `shared/components/comment/comment.component.html` file and add a card:

```
<mat-card *ngIf="comment" class="comment-card">
</mat-card>
```

Inside the card, add the card's header:

```
<mat-card-header>
</mat-card-header>
```

Inside the header, add the following markup to display the avatar of the comment's author:

```
<img mat-card-avatar class="comment-user-avatar"
  [src]="comment.author.image" />
Then, add the card's title:
<mat-card-title>
<h4>{{comment.author.fullName}}</h4>
</mat-card-title>
```

Then, add the card's subtitle with the following markup:

```
<mat-card-subtitle class="comment">
<p>
  {{comment.comment}}

  {{comment.createdAt}}
  <button mat-icon-button
   matTooltip="Remove" (click)="remove()">
   x
  </button>
</p>
</mat-card-subtitle>
```

Here, we displayed the comment, the comment's date, and the remove button for removing the comment.

Then, we applied the mat-icon-button directive to this button to create an icon button with Material styling and behavior and we used the matTooltip attribute to add a tooltip for when we hover over the button.

Now, let's go back to the template of the post component and add the following division just below the division with the comment-box class, which contains the comment's textarea element:

```
<div *ngIf="commentsShown">
  <ng-content></ng-content>
</div>
```

This slot will be used to project/insert the markup for displaying comments from the parent component right to this division.

11. Open the `users/components/profile/profile.component.ts` file and add the following method to the component's class:

```
postComments(postId: string) {
  return this.comments.get(postId)?.result!;
}
```

This will be used to access the `comments` array from the `comments` map by post ID.

12. Open the `users/components/profile/profile.component.html` file, add the handlers for the remaining custom events, and display the post's comments, as follows:

```
<div *ngFor="let post of posts;">
  <app-post [post]="post" [authUser]="authUser"
    (remove)="onRemovePost($event)"
    (comment)= "onComment($event)"
    (like)= "onLike($event)"
    (listComments)= "onListComments($event)"
    (moreComments)= "onLoadMoreComments($event)">
  </app-post>
</div>
```

Inside the `<app-post>` element, display the comments, as follows:

```
<div class="comments-section"
*ngFor="let comment of postComments(post.id)">
  <app-comment
  [comment]="comment"
  (removeComment)="onRemoveComment($event)"></
app-comment>
</div>
```

Here, we added an HTML division with the `comments-section` class, which will be projected into the `<post>` component inside the HTML division with the `<ng-content>` element. Inside this division, we iterated over the post's comments using the `*ngFor` directive and we displayed each comment using the `<app-comment>` element. We passed the `comment` object via the `comment` property and set the handler for the `removeComment` event to the `onRemoveComment()` method, which was implemented in the base component.

The following screenshot shows the comment that's displayed:

Figure 11.7 – A comment being displayed

Now, we can display the comments and load more comments. However, if we add a comment, it will not be displayed automatically until we click on one of the buttons. We'll fix this in the next section.

Displaying new comments

We can implement this either by re-fetching comments when a new comment is added or by manually adding the comment to the cache.

We can re-fetch comments by updating the createComment() method in the core/services/comment/comments.service.ts file, as follows:

```
createComment(
  comment: string,
  postId: string) {
  return this.commentPostGQL
    .mutate({
      comment: comment,
      postId: postId
    }, {
      refetchQueries: ['getCommentsByPostId']
    })
```

```
        .pipe(map(result => result.data!.comment));
}
```

However, be careful when you're re-fetching the comments since this will re-run all the `getCommentsByPostId` active queries for all the posts that we previously fetched comments for.

For example, in my case, the following screenshot shows the currently active queries in Apollo DevTools:

Figure 11.8 – Active queries in Apollo DevTools

Here, you can see that we can have multiple active `getCommentsByPostId` queries at the same time.

Now, let's learn how to manually update the cache. Update the `createComment()` method by adding the following object, along with the `update()` method, as the second argument to the `mutate()` method:

```
{
  update(cache, comment) {
    cache.modify({
      fields: {
        getCommentsByPostId(existingComments) {
          const comments = [comment, ...existingComments]
          return comments;
        }
```

```
        }
    });
  }
}
```

Next, we'll learn to update the cache after removing comments.

Updating the cache after removing comments

If we want to remove comments, we can click on the **x** button that exists next to the comment. At this point, the comment gets removed, but we need to refresh the comments to verify that it was deleted.

We want to see the result immediately. Again, we can implement this by re-fetching the comments or manually updating the cache. Update the `removeComment()` method in the `core/services/comment/comments.service.ts` file by adding a second argument object with an `update()` method, as follows:

```
{
  update(cache) {
    cache.modify({
      fields: {
        getCommentsByPostId(existingComments, { readField }) {
          return existingComments
            .filter((cc: StoreObject) => {
              return id !== readField('id', cc)
            });
        }
      }
    });
  }
}
```

We have already imported `StoreObject`, which simply represents an object in the cache. In the `update()` function, we add the code for updating the cache by removing the deleted comment using the `filter()` method and the `readField()` utility function, which is used to read the value of the cached field. For more information, check out https://www.apollographql.com/docs/react/caching/cache-interaction/#example-removing-an-item-from-a-list.

More information on this is also available at `https://www.apollographql.com/docs/react/caching/overview/`.

Giving focus to the commenting box

After displaying the comment box, we want to automatically have a focus on `<textarea>` so that we can start typing our comment immediately. We can implement this feature using a custom Angular directive, as follows:

1. Head over to your Terminal and run the following command:

    ```
    ng g directive shared/directives/autofocus
    ```

    This directive is automatically imported into the shared module by the CLI. You don't need to re-export it like you have to with the post and comment components since it will only be used inside the component(s) of the shared module.

2. Open the `shared/directives/autofocus.directive.ts` file and update it, as follows:

    ```
    import { Directive, AfterViewInit } from
      '@angular/core';
    import { MatInput } from '@angular/material/input';

    @Directive({
      selector: '[appAutofocus]'
    })
    export class AutofocusDirective implements
      AfterViewInit {

        constructor(private matInput: MatInput) { }
          ngAfterViewInit() {
          this.matInput.focus();

      }
    }
    ```

Here, we imported the `AfterViewInit` interface and the `MatInput` directive before implementing the life cycle interface and injecting the directive.

Finally, we called the `focus()` method of the `MatInput` instance inside the `ngAfterViewInit()` method, which gets called once the view has been initialized.

> **Note**
>
> See `https://angular.io/api/core/AfterViewInit` and `https://material.angular.io/components/input/overview` for more information.

Finally, in the `shared/components/post/post.component.html` file, apply the custom `appAutofocus` directive to the `<textarea>` element to automatically give it focus when it gets displayed:

```
<textarea #commentInput placeholder="Leave a
    comment.." matInput (keydown.
enter)="createComment($event)"
    appAutofocus></textarea>
```

The following screenshot shows the comment box with focus after clicking on the **Comment** button:

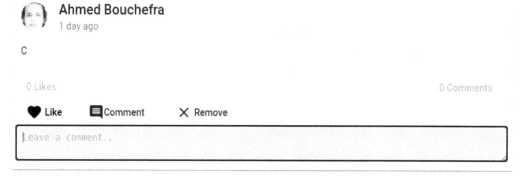

Figure 11.9 – Comment text area with focus

We finished implementing our profile component by adding the code and markup for displaying posts with comments and likes. In the same way, we need to implement the posts component of the feeds module, which is the home page where the latest posts of all the users of the network should be displayed.

Since the feed's posts component is similar to the profile component and the most common code is implemented inside the base class, which can be also extended by the feed's posts component, we'll leave implementing this to you as an exercise.

Assignments

As an assignment, implement the pagination for the users' search functionality. In *Chapter 10, Fetching Posts and Adding Comments and Likes*, we added a button for fetching more users in the search dialog. However, we didn't implement all the required code both on the server and client apps. So, as an assignment, add the necessary code to make pagination work with the user's search results.

As a second assignment, implement the `getFeed()` method of the post service to fetch the latest posts of the network. It has the following signature:

```
getFeed(offset: number, limit: number){}
```

This method is similar to the `getPostsByUserId()` method except that we don't need to pass the user ID argument. We previously defined the GraphQL query and generated the variables and the result types that were required by this method, so you simply need to write the body of the method.

You also need to add the necessary field policy to the cache for the `Query.getFeed` field.

Finally, implement the posts component of the feed module. This component should display the first name of the logged-in user and the user feed. The following screenshot shows what the component should look like after adding a user photo and some posts:

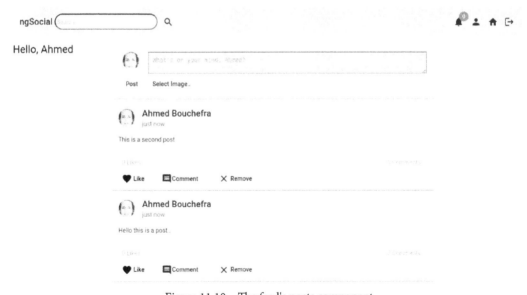

Figure 11.10 – The feed's posts component

Finally, write the necessary tests, similar to how you did in the previous chapters.

Next, let's learn how to set up Apollo Client for subscriptions.

Setting up Apollo Client for subscriptions

In this section, we'll set up Apollo Client for subscriptions to notify users if the other users on the network liked or commented on their posts in real time.

Subscriptions need a persistent connection, so they can't be handled using the default HTTP protocol that Apollo Client uses for queries and mutations. Instead, they are sent over WebSocket, via the community-maintained `subscriptions-transport-ws` library, which we need to install in our project:

1. Head over to your Terminal, make sure you have navigated inside your frontend project's folder (the `client/` folder), and run the following command:

   ```
   npm install subscriptions-transport-ws
   ```

2. Open the `src/app/graphql.module.ts` file and start by adding the following imports:

   ```
   import {
     ApolloClientOptions,
     from,
     split
   } from '@apollo/client/core';
   import { WebSocketLink } from
     '@apollo/client/link/ws';
   import { getMainDefinition } from
     '@apollo/client/utilities';
   ```

   Here, we imported the `WebSocketLink` and `getMainDefinition` symbols from their packages and added the `split` method to the existing import declaration from `@apollo/client/core`.

3. Inside the `createApollo()` function, define the following Apollo link:

   ```
   const ws = new WebSocketLink({
     uri: 'ws://localhost:8080/graphql',
     options: {
       reconnect: true,
   ```

```
        connectionParams: {
          authToken: accessToken
        }
      }
    });
```

Here, we created a WebSocket link to the `ws://localhost:8080/graphql` URL where the WebSocket server is listening. We specified some options, such as setting the `reconnect` attribute to `true` to reconnect if the WebSocket is disconnected, and using the `connectionParams` object to pass the authentication token to the server.

4. In the same function, add the following code to split the links, as follows:

```
    const splitLink = split(
      ({ query }) => {
        const definition = getMainDefinition(query);
        return (
          definition.kind === 'OperationDefinition' &&
          definition.operation === 'subscription'
        );
      },
      ws,
      http
    );
```

Here, we used the `split()` function to split the links so that we can send the appropriate data to each link, depending on the type of operation that is being sent – a query, a mutation, or a subscription.

5. Finally, replace the HTTP link with the split link in the chain of links that are returned from the `createApollo()` function:

```
    return {
      link: from([setAuthorizationLink, splitLink]),
      cache: cache
    };
```

Now, we have a link to both the HTTP and WebSocket servers that are running on different ports. Check out the source code commit from `https://git.io/JDuwf`.

Next, we'll learn how to implement realtime subscriptions in our client application.

Implementing GraphQL subscriptions

Now that we've set up Apollo Client for GraphQL subscriptions, we'll learn how to use subscriptions to add some realtime features to our application.

In the server, we have added a `Subscription` type with two fields – `onPostCommented` and `onPostLiked` – to our GraphQL schema. Then, we implemented the required resolvers to push realtime updates to the client when a post is commented on or liked.

In this section, we are going to learn how to listen to realtime updates from the server. This process is similar to how we used queries to query for data, but instead of using the query operation, we use the subscription operation with the fields that need to be sent from the server. For the response, we get results every time new data is available on the server.

In our case, we'll implement the functionality for displaying new comments and likes on posts every time they are added by the users of the network without the need to refresh the page or re-click buttons. We'll also implement a simple notification system that notifies the users of new comments and likes.

Before we proceed, we need to fix a line of code in the server. Open the `server/src/index.ts` file and locate the `SubscriptionServer.create()` method. Then, update the code for retrieving the token from the `connectionParams` object inside the `onConnect` handler, as follows:

```
const token = connectionParams['authToken'] || '';
```

Check out the changes by reading the following code commit: `https://git.io/JDuom`.

Let's start by adding the subscription for listening to new comments on posts:

1. Open the `src/app/graphql.documents/comment.graphql` file and add the following subscription:

    ```
    subscription onPostCommented{
      onPostCommented {
        ...CommentFields
        author { ...BasicUserFields }
        post { id commentsCount }
      }
    }
    ```

The `onPostCommented` subscription returns a `Comment` object. Due to this, we must use a selection set to ask for the fields that we are interested in, such as the ID, the comment's text, the author's information, the ID of the commented post, and its creation time.

2. Open the `src/app/graphql.documents/like.graphql` file and add the following subscription to listen for new likes:

```
subscription onPostLiked{
  onPostLiked{
    id
    user { ...BasicUserFields }
    post { id likesCount }
    createdAt
  }
}
```

3. After adding these subscriptions, run the code generator from the root of your monorepo project using the following command:

```
npm run codegen
```

4. In the `src/app/core/services/comment/comments.service.ts` file, add the following import:

```
import { OnPostCommentedDocument }
  from '@ngsocial/graphql/documents';
```

5. Update the `getCommentsByPostId()` method by adding the following code:

```
queryRef.subscribeToMore({
  document: OnPostCommentedDocument,
  updateQuery: (prev, { subscriptionData }) => {}
});
```

Here, we used `subscribeToMore()` of the `QueryResult` object that's returned by the `watch()` method to subscribe to realtime updates from the `onPostCommented` subscription. We called the `subscribeToMore()` method of the `queryRef` variable and set the `document` attribute of the argument object to the `onPostCommentedDocument` subscription.

6. In the arrow function that's passed to the `updateQuery` property, add the following code:

```
if (!subscriptionData.data) {
  return prev;
}
const newComment = subscriptionData
  .data
  .onPostCommented!;
console.log("New comment coming: ",
  subscriptionData.data);
if (newComment.post.id !== postId) {
  return prev;
}
return {
  getCommentsByPostId: [newComment,
    ...prev.getCommentsByPostId!]
};
```

Here, we implemented the `updateQuery` callback function, which gets called every time we get new data from the server, to merge the new comment with the existing comments. This function should return an object that has the shape of the initial query data.

7. Open the `src/app/core/services/like/likes.service.ts` file and add the following import:

```
import { OnPostLikedDocument }
  from '@ngsocial/graphql/documents';
```

8. Update the `getLikesByPostId()` method, as follows:

```
queryRef.subscribeToMore({
  document: OnPostLikedDocument,
  updateQuery: (prev, { subscriptionData }) => {}
});
```

9. In the function that's passed to the `updateQuery` property, add the following code:

```
if (!subscriptionData.data) {
  return prev;
}
const newLike = subscriptionData
  .data
  .onPostLiked!;
console.log("New like coming: ",
  subscriptionData.data);
if (newLike.post.id !== postId) {
  return prev;
}
return {
  getLikesByPostId: [newLike,
    ...prev.getLikesByPostId!]
};
```

Now, if someone adds a comment to a post, it will be added and shown in real time in the comments area. But before you start getting realtime updates of new comments, the user needs to run the `getCommentsByPostId()` method at least once by clicking on one of the two buttons on the post that fetches the comments. Check out the source code commit by going to `https://git.io/JDurx`.

If we need to implement separate logic, such as for showing notifications about new comments before we fetch them, we must use the GraphQL services that extend the `Subscription` service that we generated previously:

1. Open the `src/app/core/components/header/header.component.ts` file and import the following generated services:

```
import {
  OnPostCommentedGQL,
  OnPostLikedGQL
} from '../../gql.services';
```

2. Define the following property in the component's class:

```
public notificationsCount: number = 0;
```

This will be used to display the notification's number on the header's template.

3. Inject the services that you imported via the service's constructor:

```
constructor(
  // [...]
    private onPostCommentedGQL: OnPostCommentedGQL,
    private onPostLiked: OnPostLikedGQL) { }
```

4. Add the following method for subscribing to new comments to the component's class:

```
subscribeToNewComments() {
  const onPostCommentedObs = this.onPostCommentedGQL
    .subscribe();
  onPostCommentedObs
    .subscribe({
      next: (result) => {
        const comment = result.data?.onPostCommented;
        if (this.authUser?.id !== comment?.author.id) {
          this.notificationsCount++;
          this.changeDetectorRef.markForCheck();
        }
      }
    });
}
```

Here, we called the `subscribe()` method of the GraphQL service to return an Observable that we can subscribe to, to start listening for new comments.

Then, we subscribed to the returned Observable to start listening for new comments. In the next handler of the observer object, we retrieved the comment from the GraphQL subscription response; we increment the `notificationsCount` property if the comment's author is different from the currently logged-in user.

5. Similarly, add the following method for subscribing to new likes to the component's class:

```
subscribeToNewLikes() {
  const onPostLikedObs = this.onPostLiked
    .subscribe();
  onPostLikedObs.subscribe({
    next: (result) => {
      const like = result.data?.onPostLiked;
      if (this.authUser?.id !== like?.user.id) {
        this.notificationsCount++;
        this.changeDetectorRef.markForCheck();
      }
    }
  });
}
```

6. Call the previous methods from the ngOnInit() method of the component, as follows:

```
ngOnInit(): void {
  this.authService.authState
  .pipe(takeUntil(this.destroyNotifier$))
  .subscribe({
    next: (authState: AuthState) => {
      this.isLoggedIn = authState.isLoggedIn;
      this.authUser = authState.currentUser;
      this.changeDetectorRef.markForCheck();
      this.subscribeToNewComments();
      this.subscribeToNewLikes();
    }
  });
}
```

7. Finally, display the notifications count in the template using a Material badge. Open the `src/app/core/components/header/header.component.html` file and update the **Notifications** button, as follows:

```
<button *ngIf="isLoggedIn" mat-icon-button
matTooltip="Notifications">
    <mat-icon aria-hidden="false" aria-
  label="Notifications" [matBadge]=
    "notificationsCount"matBadgeColor="warn">
      notifications
    </mat-icon>
</button>
```

Here, we added the `matBage` attribute to the `<mat-icon>` element and bound it to the `noticationsCount` property of the component.

That's it – you should now be able to receive and display the notifications of new comments and likes in your application. Check out the commit at `https://git.io/JDuoK`.

You can always extend this implementation to display more information about the notifications the users receive.

Summary

In this final chapter, we learned how to configure Apollo Client for GraphQL subscriptions and then added code to display notifications in the application's header using a badge.

Throughout this book, we have used cutting-edge technologies such as Angular and GraphQL to develop a full stack application with a monorepo architecture using Lerna. We used Apollo on both the client and server to send and respond to GraphQL queries.

After setting up the development environment and installing Node.js, we built a server with GraphQL support to implement the backend using Express.js and Apollo Server. We also used Apollo Studio to communicate with our GraphQL API before we started developing a frontend application to consume the API and present a user interface to users to communicate with the backend.

After that, we explained how to install Express.js and configure it with TypeScript and GraphQL. We used mocking to provide a working GraphQL server with Apollo Server before implementing the resolvers that are responsible for fetching and adding data.

Then, we learned how to connect a MySQL database to our web application using TypeORM and implemented the resolvers for communicating with the database. We exposed a working GraphQL API with resolvers that query and remove data from a real MySQL database.

We also utilized TypeORM to abstract the database operations, allowing developers to use any desired database management system for their application without changing the code. We learned how to connect TypeORM to Apollo, as well as how to generate and seed database tables.

Next, we added authentication and image uploads with Apollo Server to our GraphQL API and learned about the necessary concepts for adding authentication with Node.js, Express, and Apollo Server, as well as for handling image uploads.

We also added realtime support with authentication to our server application, which enabled us to communicate fresh data from the server to the client as soon as it becomes available. To do this, we leveraged Apollo Server's GraphQL subscriptions to push new comments and likes on users' posts from the server to the client, right at the moment when they are added.

After implementing the backend, we used the Angular CLI to create a new Angular project and reviewed many fundamental Angular concepts, such as modules, components, and services. We also learned how to use the Angular CLI to generate the artifacts that constitute the structure of the application.

Then, we utilized Angular Material in the project to provide Material design components to create a visually appealing application UI.

Finally, we integrated the frontend that we built using Angular with the backend using Apollo Client, which is designed to send GraphQL queries and mutations to the server to fetch and write data.

After developing and testing the application and adding any extra features that you want, you can build the production bundles and host them with your preferred hosting service. Please refer to the official documentation at `https://angular.io/guide/deployment` for more information about this process.

Thanks for taking the time to read this book and I hope you enjoyed it.

Index

U

W

Packt.com

Subscribe to our online digital library for full access to over 7,000 books and videos, as well as industry leading tools to help you plan your personal development and advance your career. For more information, please visit our website.

Why subscribe?

- Spend less time learning and more time coding with practical eBooks and Videos from over 4,000 industry professionals

- Improve your learning with Skill Plans built especially for you

- Get a free eBook or video every month

- Fully searchable for easy access to vital information

- Copy and paste, print, and bookmark content

Did you know that Packt offers eBook versions of every book published, with PDF and ePub files available? You can upgrade to the eBook version at packt.com and as a print book customer, you are entitled to a discount on the eBook copy. Get in touch with us at customercare@packtpub.com for more details.

At www.packt.com, you can also read a collection of free technical articles, sign up for a range of free newsletters, and receive exclusive discounts and offers on Packt books and eBooks.

Other Books You May Enjoy

If you enjoyed this book, you may be interested in these other books by Packt:

Accelerating Angular Development with Ivy

Lars Gyrup Brink Nielsen | Jacob Andresen

ISBN: 978-1-80020-521-5

Find out why Angular Ivy tests are faster and more robust.

- Explore the concept of CSS custom properties and scoping of values and learn how to use them with Angular Ivy.

- Use testing harnesses present in Angular components to write effective tests.

- Explore the architecture of the Angular compatibility compiler and understand why it is important.

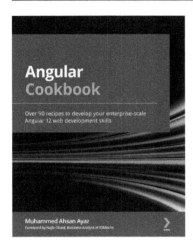

Angular Cookbook

Muhammad Ahsan Ayaz

ISBN: 978-1-83898-943-9

- Gain a better understanding of how components, services, and directives work in Angular.

- Understand how to create Progressive Web Apps using Angular from scratch.

- Build rich animations and add them to your Angular apps.

- Manage your app's data reactivity using RxJS.

- Implement state management for your Angular apps with NgRx.

Packt is searching for authors like you

If you're interested in becoming an author for Packt, please visit authors. packtpub.com and apply today. We have worked with thousands of developers and tech professionals, just like you, to help them share their insight with the global tech community. You can make a general application, apply for a specific hot topic that we are recruiting an author for, or submit your own idea.

Share Your Thoughts

Now you've finished *Full-Stack Development with Angular and GraphQL*, we'd love to hear your thoughts! Scan the QR code below to go straight to the Amazon review page for this book and share your feedback or leave a review on the site that you purchased it from.

https://www.amazon.in/review/create-review/error?asin=1800202466

Your review is important to us and the tech community and will help us make sure we're delivering excellent quality content.

www.ingramcontent.com/pod-product-compliance
Lightning Source LLC
Chambersburg PA
CBHW062039050326
40690CB00016B/2980